Engineering Dynamics

Synthesis Lectures on Mechanical Engineering

Synthesis Lectures on Mechanical Engineering series publishes 60–150 page publications pertaining to this diverse discipline of mechanical engineering. The series presents Lectures written for an audience of researchers, industry engineers, undergraduate and graduate students.

Additional Synthesis series will be developed covering key areas within mechanical engineering.

Engineering Dynamics
Cho W.S. To
2018

Solving Practical Engineering Problems in Engineering Mechanics: Dynamics
Sayavur I. Bakhtiyarov
2018

Solving Practical Engineering Mechanics Problems: Kinematics
Sayavur I. Bakhtiyarov
2018

C Programming and Numerical Analysis: An Introduction
Seiichi Nomura
2018

Mathematical Magnetohydrodynamics
Nikolas Xiros
2018

Design Engineering Journey
Ramana M. Pidaparti
2018

Introduction to Kinematics and Dynamics of Machinery
Cho W. S. To
2017

Microcontroller Education: Do it Yourself, Reinvent the Wheel, Code to Learn
Dimosthenis E. Bolanakis
2017

Solving Practical Engineering Mechanics Problems: Statics
Sayavur I. Bakhtiyarov
2017

Unmanned Aircraft Design: A Review of Fundamentals
Mohammad Sadraey
2017

Introduction to Refrigeration and Air Conditioning Systems: Theory and Applications
Allan Kirkpatrick
2017

Resistance Spot Welding: Fundamentals and Applications for the Automotive Industry
Menachem Kimchi and David H. Phillips
2017

MEMS Barometers Toward Vertical Position Detecton: Background Theory, System
Prototyping, and Measurement Analysis
Dimosthenis E. Bolanakis
2017

Engineering Finite Element Analysis
Ramana M. Pidaparti
2017

Engineering Dynamics
Cho W.S. To

ISBN: 978-3-031-79616-6   paperback
ISBN: 978-3-031-79617-3   ebook
ISBN: 978-3-031-79618-0   hardcover

DOI 10.1007/978-3-031-79617-3

A Publication in the Springer series
SYNTHESIS LECTURES ON MECHANICAL ENGINEERING

Lecture #15
Series ISSN
Print 2573-3168   Electronic 2573-3176

Engineering Dynamics

Cho W.S. To
University of Nebraska, Lincoln

SYNTHESIS LECTURES ON MECHANICAL ENGINEERING #15

ABSTRACT

Engineering Dynamics is an introductory textbook covering the kinematics and dynamics of particles, systems of particles, and kinematics and dynamics of rigid bodies. It has been developed from lecture notes given by the author since 1982. It includes sufficient topics normally covered in a single-semester three credit hour course taken by sophomores in an undergraduate degree program majoring in various engineering disciplines.

The primary focus of the book is on kinematics and dynamics of particles, kinematics and dynamics of systems of particles, and kinematics and dynamics of rigid bodies in two- and three-dimensional spaces. It aims at providing a short book, relative to many available in literature, but with detailed solutions to representative examples. Exercise questions are included.

KEYWORDS

kinematics, dynamics, particles, systems of particles, rigid bodies

Contents

Preface

This book is concerned with discrete engineering dynamics in the sense that particles, systems of particles, rigid bodies in two-dimensional (2D) and three-dimensional (3D) spaces are included. Non-discrete engineering dynamics, which includes dynamic problems of elastic bodies or continuous or distributed parameter models, is outside the scope of this book and therefore it is not included. The exclusion of dynamic problems of elastic bodies is due to two main considerations. First, the present book is aimed at undergraduate sophomore students pursuing an engineering degree. Second, dynamic problems of elastic bodies generally involve with more mathematical tools and techniques which are well beyond the background preparation of sophomore students.

There are various excellent books in *Engineering Dynamics* at a similar level. However, the lengths of these books are generally more than 400–500 pages and it seems that there is a lack of shorter books in the range of 150 and 200 pages. While the present book is much shorter in length compared with existing textbooks available in the literature, it does include (a) topics normally covered in a three-credit sophomore undergraduate course, and (b) many examples are solved with detailed steps throughout. Owing to the limited length and the author's belief that it is more important to understand the issues, concepts, or theory and how the steps in the solution of a particular problem are applied than attempting to work on as many problems as one can, exercises at the end of every chapter are selective and representative. The author agrees that "practice makes perfect" and therefore many exercise problems should be attempted, if time permits. He strongly believes, however, that such a learning process can often lead to develop an ability to problems solved by "pattern recognition". Besides, many students are overwhelmed by homework assignments every week and therefore it is unrealistic to expect them to devote too much time on any one subject. This is why a book of shorter length that emphasizes on understanding and learning the steps in the solution for a particular problem is more important and efficient.

This book is based on the lecture notes that have been developed and used by the author since 1982. These lecture notes have been employed in the titled course first taught at The University of Calgary, Canada and subsequently at the University of Nebraska, Lincoln, Nebraska.

Under normal conditions, it is expected that the students using the present book have already taken a second course in engineering mathematics and engineering statics in an undergraduate degree program in engineering.

Cho W.S. To
May 2018

Acknowledgments

The author would like to express his sincere thanks to Paul Petralia, Executive Editor, Engineering, and his team members at Morgan & Claypool for their assistance and production effort on this book.

Cho W.S. To
May 2018

CHAPTER 1

Introduction

1.1 BACKGROUND AND MOTIVATION

Engineering dynamics is an extremely interesting subject and is rich in topics that are important to the solution of everyday problems encountered in the design and analysis in engineering. There are many interesting examples in the conceptual and physical worlds. In this chapter only four *classical* examples are presented to illustrate the interesting and useful aspects of engineering dynamics. These examples, while regarded as classical (non-quantum), have many relevant modern applications and important results that remain to be explored and deep understanding to be made. In the following sections these examples are to be included accordingly.

1.1.1 CLASSICAL WATER MOLECULE AND OZONE MOLECULE

The water molecule is composed of three particles interacting by interparticle *conservative* forces, as illustrated in Figure 1.1 in which m_o is the mass of the oxygen atom and m the mass of the hydrogen atom. When $m_o = m$, the system becomes the ozone molecule that is of great interest to environmental engineers and scientists. The water molecule is of course the classical *three-body problem* in space [1, 2]. It is a special case of the *n-body problem*. For the three-body problem there is no general *closed-form solution* for every condition.

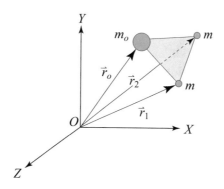

Figure 1.1: Water molecule in 3D space.

Historically, the first specific three-body problem that received extended study was that dealt with the Moon, the Earth, and the Sun [1]. The gravitational problem of three bodies in space dates from 1687, when Newton [3] published his famous *Principia*.

1.1.2   NEWTON'S CRADLE

The term *Newton's cradle* is used mainly for terminological convenience and for commercial iden-
tification in toy shops. A commonly available version is that shown in Figure 1.2 in which five
identical spherical metal balls are similarly suspended by wires attached to both sides. The main
technical interest of this toy is in its use to demonstrate the impact of multiple particles. For example, if one pulls two spherical balls on the left-hand side (lhs) of the cradle and releases them
simultaneously so that they hit the three remaining stationary balls. It is interesting to observe
that only two of the three stationary balls will fly to the right-hand side (rhs). If one pulls three
balls to the lhs and releases them so that they hit the two remaining balls. It is observed that
only the three balls will fly to the rhs.

Figure 1.2: Newton's cradle.

1.1.3   SIMPLE TOY ROTOR

The simple rotor can easily be found in many toy shops. The bottom view of one version is shown
in Figure 1.3a, while the sketch of side view of the toy rotor on a glass panel is included in Figure 1.3b where the counter-clockwise black arrows denote the direction of the applied transient
torque (produced by the twisting action from the holding fingers that are quickly removed). The
transient torque enables the rotor to rotate in the counter-clockwise direction for 2 or 3 s (the
duration of this rotation depends on the magnitude of the twisting action) before it stops momentarily for a fraction of a second. Then the rotor rotates freely in the clockwise direction (that
is, the direction of rotation now is opposite to that of the originally applied transient torque).
This rotation is indicated by the two blue arrows in the figure. The interesting point to note in
this system is that the rotor rotates in one direction and stops momentarily before it rotates freely
in the opposite direction. This example demonstrates the importance of the distribution of mass
moment of inertia on stability in the design of propellers in ships, aircrafts, and helicopters.

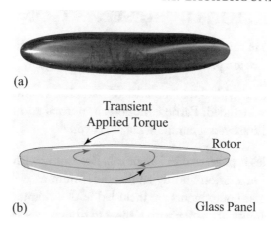

Figure 1.3: (a) Bottom view of a toy rotor and (b) side view of toy rotor.

1.1.4   FALLING CAT PROBLEM

The *falling cat problem* is concerned with the understanding of the physics and explanation of how a falling cat can change its orientation such that it is able to land its feet on the ground, irrespective of its initial orientation, and without violating the law of *conservation of angular momentum*. Photographs of a falling cat taken with a high-speed camera [4] are included in Figure 1.4 in which the sequential motion of a falling cat is from (a) through (d).

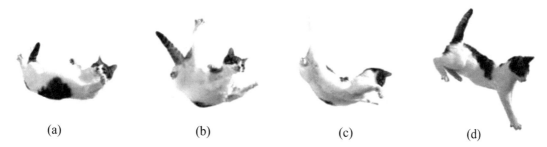

Figure 1.4: Sequential motion of a falling cat.

The falling cat problem has drawn interest from famous scientists including Maxwell [5]. The solution of the problem is due to Kane and Scher [6] in which the cat was simplified as a pair of coupled cylinders (the front and back halves of the cat) that can change their relative orientations. Montgomery later [2, 7] considered the Kane-Scher model in terms of a connection in the configuration space. One application of the results of such a study is in the *orientation* or so-called *attitude control* of satellites in orbit.

1.2 ORGANIZATION OF PRESENTATION

The original lectures given in this course, *Engineerig Dynamics*, are organized into eight chapters. Thus, the remaining lectures are presented in Chapters 2 through 8.

- Chapter 2 is concerned with the kinematics of particles. Rectilinear and curvilinear motions of particles are included. Particle motions in normal and tangential components as well as in radial and transversal components are covered.

- Dynamics of particles is presented in Chapter 3. Beginning with Newton's laws of motion topics on equations of motion and dynamic equilibrium as well as equations of motion in radial and transverse components are included in this chapter. Linear and angular momenta as well as their rate of change with respect to time are also considered in this chapter.

- Work and energy of particles are presented in Chapter 4. Topics covered in this chapter include potential energy and strain energy, kinetic energy of a particle, principle of work and energy, and principle of conservation of energy.

- Chapter 5 is involved with impulse, momentum as well as impact of particles. The principle of impulse and momentum, impulsive motion and impact, direct central impact as well as oblique central impact, and constrained oblique central impact are covered in this chapter.

- Systems of particles are considered in Chapter 6 in which Newton's laws of motion for systems of particles are covered. Linear and angular momentum of a system of particles are discussed. Motion of mass center of a system of particles is presented along with angular momentum of a system of particles are included. Conservation of momentum and work energy principle for a system of particles are also presented in this chapter.

- Chapter 7 is concerned with the kinematics of rigid bodies. Three-dimensional motion of rigid bodies or links with respect to a rotating frame of reference (*RFR*) is emphasized in this chapter.

- Chapter 8, the final chapter, is concerned with the dynamics of rigid bodies in 2D and 3D spaces. Equations of motion for rigid bodies, linear and rotational, are presented. The principle of work and energy as well as the conservation of energy and conservation of momentum, and principle of impulse and momentum are included in this chapter. Eulerian angles and motion of a gyroscope are also introduced.

REFERENCES

[1] Poincaré, H., *New Methods of Celestial Mechanics*, volumes I–III, American Institute of Physics, 1967. 1

[2] Marsden, J. E., Lectures on mechanics, *London Mathematical Society Lecture Note Series 174*, Cambridge University Press, 1992. DOI: 10.1017/cbo9780511624001. 1, 3

[3] Newton, I., *Philosophiæ Naturalis Principia Mathematica*, 1687. DOI: 10.5479/sil.52126.39088015628399. 1

[4] https://www.bing.com/images/search?q=falling+cat+photos&id 3

[5] Campbell, L. and Garnett, W., *The Life of James Clerk Maxwell*, Macmillan and Company, 1992. 3

[6] Kane, T. R. and Scher, M. P., A dynamical explanation of the falling cat phenomenon, *International Journal of Solids and Structures*, 5, pp. 663–670, 1969. DOI: 10.1016/0020-7683(69)90086-9. 3

[7] Montgomery, R., Gauge theory of the falling cat, in Enos, M. J., *Dynamics and Control of Mechanical Systems*, American Mathematical Society, pp. 193–218, 1993. 3

CHAPTER 2

Kinematics of Particles

2.1 INTRODUCTION

Kinematics of particles has to do with the studies of linear and angular position, velocity, and acceleration of particles. The question of why one has to study these quantities in the context of engineering is best answered by the following examples. In rotating machinery the *position* or *displacement* is required in the design criterion or engineering specifications. The displacement is used to qualify the smoothness or roughness of the motion, for example. *Velocity* is, however, related to the *kinetic energy* of the particle and kinetic energy is frequently used in the study of impact or collision which, in turn, is required in the study of *fracture*. *Acceleration* is related to dynamic *force* since by definition dynamic force is a product of mass and acceleration. In other words, position, velocity, and acceleration are the important and basic concepts that have to be studied first in the context of engineering.

In the next section, rectilinear motion of particles is introduced. Uniform and uniformly accelerated rectilinear motion is treated in Section 2.3. Curvilinear motion of particles is considered in Section 2.4. Motion in rectangular components is dealt with in Section 2.5 while motion in tangential and normal is introduced in Section 2.6. Motion in radial and transverse components is presented in Section 2.7.

2.2 RECTILINEAR MOTION OF PARTICLES

Rectilinear motion of a particle is concerned with the motion of a particle along a straight line. The concept of a straight line naturally leads to the definition of the position of a particle. This, in turn, needs the point of reference and such a point can be imagined as a fixed point along the straight line. For illustration, one can call this fixed point O, as shown in Figure 2.1. The position of the particle is at x measuring from the fixed point O along the straight line OxX which is called the *position coordinate*.

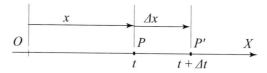

Figure 2.1: Illustration of particle position.

The instantaneous velocity is defined as

$$v = \lim_{\Delta t \to 0} \frac{\Delta x}{\Delta t} = \frac{dx}{dt}. \tag{2.1}$$

The instantaneous acceleration is defined as

$$a = \lim_{\Delta t \to 0} \frac{\Delta v}{\Delta t} = \frac{dv}{dt} = \frac{d^2x}{dt^2}. \tag{2.2}$$

This acceleration can also be written as

$$a = \frac{dv}{dt} = \frac{dv}{dx}\frac{dx}{dt} = \frac{dv}{dx}v. \tag{2.3}$$

Example 2.1
In a series of experiments the motion of a particle is found to satisfy the equation $x = t^3 - 6t^2 + 9t + 15$, where x and t are in m and s, respectively. Find the time, position, and acceleration when the velocity $v = 0$.

Solution:
Given equation for displacement is $x = t^3 - 6t^2 + 9t + 15$.
Differentiating with respect to (w.r.t.) time t once and twice,

$$v = \frac{dx}{dt} = \frac{d\left(t^3 - 6t^2 + 9t + 15\right)}{dt} = 3t^2 - 12t + 9 = 3\left(t^2 - 4t + 3\right)$$
$$= 3(t-1)(t-3).$$
$$a = \frac{dv}{dt} = 6t - 12.$$

Therefore, when $v = 0$, one obtains $t = 1$ s and $t = 3$ s.
At $t = 1$ s,
$$x = t^3 - 6t^2 + 9t + 15 = 1^3 - 6(1)^2 + 9(1) + 15 = 19,$$
$$a = 6(1) - 12 = -6.$$

Therefore, at $t = 1$ s, $x = 19$ m, $a = -6$ m/s^2.
At $t = 3$ s,
$$x = 3^3 - 6(3)^2 + 9(3) + 15 = 15,$$
$$a = 6(3) - 12 = 6.$$

Therefore, at $t = 3$ s, $x = 15$ m, $a = 6$ m/s^2.

Example 2.2

In a large-scale laboratory investigation it was found that a particle oscillates in a friction-less strict cylindrical tube between the points $x = 4$ m and $x = 16$ m with an acceleration $a = \beta(100 - x)$, where a and x are measured in m/s^2 and m, respectively, while β is a constant. It was observed that when the particle was at $x = 100$ m its velocity was 18 m/s. It was also observed that the particle was at both $x = 40$ m and $x = 160$ m when its velocity was zero. Find (a) the value of the constant β, and (b) the velocity of the particle when $x = 120$ m.

Solution:

Given equation for acceleration is $a = \beta(100 - x)$.

Since in this case no information of time t was given and therefore Equation (2.3) should be applied. Thus,

$$a = v\frac{dv}{dx} = \beta(100 - x).$$

At $x = 40$ m, $v = 0$,

$$\int_{0}^{v} v\, dv = \int_{40}^{x} \beta(100 - x)\, dx$$

$$\frac{1}{2}v^2 = \beta\left[100x - \frac{1}{2}x^2\right]_{40}^{x} = \beta\left(100x - \frac{1}{2}x^2 - 3200\right).$$

At $x = 100$ m, $v = 18$ m/s,

$$\frac{1}{2}18^2 = \beta\left(100^2 - \frac{1}{2}100^2 - 3200\right) = \beta(1800).$$

(a) Therefore, $\beta = 0.09$ s^2.

(b) At $x = 120$ m,

$$\frac{1}{2}v^2 = 0.09\left(100.120 - \frac{1}{2}120^2 - 3200\right) = 9\left(120 - \frac{1}{2}12^2 - 32\right).$$

Therefore, $v^2 = 288$ (m/s)^2 $\Rightarrow$ 16.9706 m/s.

When two particles A and B travel along the same straight line as shown in Figure 2.2, one frequently is interested in the *relative motion* of B with respect to (w.r.t.) A.

The *relative position coordinate* of B w.r.t. A can be written as

$$x_B = x_A + x_{B/A}, \tag{2.4}$$

where the subscript A denotes particle A, and so on.

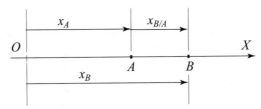

Figure 2.2: Relative motion of particles A and B.

Differentiating Equation (2.4) w.r.t. t, one has the *relative velocity* of particle B as

$$v_B = v_A + v_{B/A}. \tag{2.5}$$

Similarly, differentiating Equation (2.5) w.r.t. t, one has the *relative acceleration* of particle B as

$$a_B = a_A + a_{B/A}. \tag{2.6}$$

Example 2.3

Two cars A and B have velocities $v_A = 66$ km/h and $v_B = 48$ km/h, as shown in Figure 2.3. If the speed of each car is constant and car B arrives at the intersection 10 min after car A passed through the same intersection, find

(a) the relative velocity of B relative to A, $v_{B/A}$, and

(b) the distance between the fronts of the cars 3 min after car A passed through the intersection.

Solution:

(a) In this problem Equation (2.5) is employed. Note that the two cars are at different paths therefore Equation (2.5) should be written in vector form. That is,

$$\vec{v}_B = \vec{v}_A + \vec{v}_{B/A}.$$

The graphical representation for the present case is shown in Figure 2.3b. Thus, by using the law of cosines, one has

$$v_{B/A}^2 = 66^2 + 48^2 - 2(66)(48) \cos 155°$$

or

$$v_{B/A} = 111.366 \text{ km/h}.$$

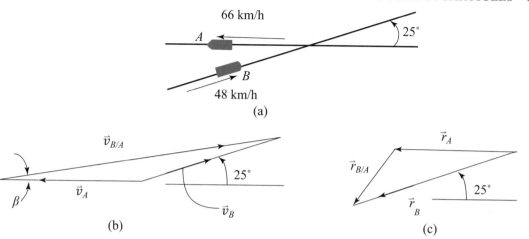

Figure 2.3: (a) Two cars at different paths, (b) graphical representation of velocities, and (c) graphical representation of position vectors.

By applying the law of sines, one obtains

$$\frac{48}{\sin \beta} = \frac{111.366}{\sin 155°}.$$

This gives

$$\beta = 10.5°.$$

Therefore,

$$\vec{v}_{B/A} = 111.366 \text{ km/h} \nearrow 10.5°.$$

(b) For this part, the relative position vector is similar to that of Equation (2.4). Again, the two cars are at different paths therefore, Equation (2.4) should be written in vector form as

$$\vec{r}_B = \vec{r}_A + \vec{r}_{B/A}.$$

At 3 min after car A passes through the intersection and the position of car A is (66 km/h) $\left(\frac{3}{60}\right) = 3.3$ km left of the intersection, while at this time the position of car B is (48 km/h) $\left(\frac{10-3}{60}\right) = 5.6$ km along the direction shown in Figure 2.3c.

With reference to Figure 2.3c, at $t = 3$ min, and applying the law of cosines, one has

$$r_{B/A}^2 = 3.3^2 + 5.6^2 - 2(3.3)(5.6) \cos 25° \quad \Rightarrow \quad r_{B/A} = 2.96 \text{ km}.$$

Therefore, the distance between the fronts of the cars 3 min after car A passed through the intersection is $r_{B/A} = 2.96$ km.

2.3 UNIFORM AND UNIFORMLY ACCELERATED RECTILINEAR MOTION

Two types of rectilinear motion are often met in the physical world. They are the *uniform rectilinear motion* and the *uniformly accelerated rectilinear motion*.

In the *uniform rectilinear motion* the velocity v of the particle is constant

$$\frac{dx}{dt} = v = constant$$

such that

$$dx = v\, dt$$

and upon integrating w.r.t. t, one has

$$x - x_o = vt \quad \text{or} \quad x = x_o + vt \tag{2.7}$$

in which the subscript denotes the initial condition.

In the *uniformly accelerated rectilinear motion* the acceleration a of the particle is constant such that upon integrating w.r.t. t once as in the following:

$$\frac{dv}{dt} = a = constant \quad \text{or} \quad dv = a\, dt$$

$$v - v_o = at \quad \text{or} \quad v = v_o + at \tag{2.8}$$

and upon integrating w.r.t. t again (starting from $\frac{dx}{dt} = v$ such that $dx = v\, dt$)

$$\int_{x_o}^{x} dx = \int_{0}^{t} v\, dt = \int_{0}^{t} (v_o + at)\, dt = v_o t + \frac{1}{2}at^2$$

to give

$$x = x_o + v_o t + \frac{1}{2}at^2. \tag{2.9}$$

If Equation (2.3) is applied,

$$\frac{dv}{dx}v = a = constant$$

such that

$$v\, dv = a\, dx$$

and upon integrating as well as rearranging, one has

$$v^2 = v_o^2 + 2a\,(x - x_o). \tag{2.10}$$

Equations (2.8), (2.9), and (2.10) are the basic equations for uniformly accelerated rectilinear motion.

The concept of relative position coordinate of two or more particles may be applied to cases in which particles or blocks (conceptually viewed as particles) are *connected by inextensible cords or cables*. For example, the blocks shown in Figure 2.4 has the *linear relation* between their position coordinates as

$$x_A + 2x_B = constant. \qquad (2.11)$$

This is known as the *geometric* or *holonomic constraint* in more advanced textbook. First and second derivatives w.r.t. t can be performed to obtain the relative velocity and acceleration.

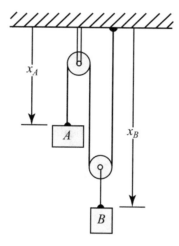

Figure 2.4: Blocks connected by inextensible cables.

Remarks:

In choosing the distances between points or blocks (conceptually regarded as points here) in this type of problems a question is often asked: Why the radius of the wheel or pulley or length of the arm that is fixed to the ground or reference positions is disregarded? The answer is that since these radius and fixtures do not change with time and therefore they can be considered as part of the constant being used in the expression or equation of position.

Example 2.4

Slider block A travels to the left, in Figure 2.5, with a constant velocity of 8 m/s and is connected by an inextensible cable to block B. It is assumed that no friction is present between the block and the supporting horizontal surface, and between the cable and pulley. Determine

(a) the velocity of block B,

(b) the velocity of portion D of the cable, and

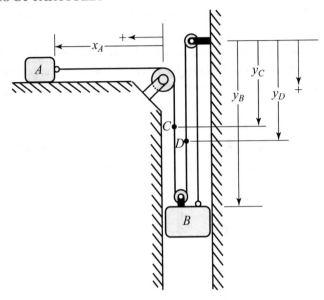

Figure 2.5: Blocks travelling horizontally and vertically.

(c) the relative velocity of portion C of the cable with respect to portion D.

Solution:
Let x and y be the horizontal and vertical distances, respectively. Assuming positive distances as shown in Figure 2.5, one has

$$x_A + 3y_B = constant.$$

Then, the velocity equation, by taking the time derivative of this position equation, becomes

$$v_A + 3v_B = 0, \qquad\qquad (2.12a)$$

taking the time derivative again, one obtains

$$a_A + 3a_B = 0. \qquad\qquad (2.12b)$$

(a) Substituting $v_A = 8$ m/s into Equation (2.12a), one has

$$v_B = -\frac{8}{3} \text{ m/s} \quad \text{or} \quad \vec{v}_B = \frac{8}{3} \text{ m/s} \uparrow .$$

(b) With reference to Figure 2.5,

$$y_B + y_D = constant$$

such that

$$v_B + v_D = 0.$$

This gives

$$v_D = \frac{8}{3} \text{ m/s} \quad \text{or} \quad \vec{v}_D = \frac{8}{3} \text{ m/s} \downarrow .$$

(c) With reference to Figure 2.5,

$$x_A + y_C = constant$$

such that

$$v_A + v_C = 0 \quad \Rightarrow \quad v_C = -v_A = -8 \text{ m/s}.$$

This gives the relative velocity

$$v_{C/D} = v_C - v_D = \left(-8 - \frac{8}{3}\right) \text{ m/s} \quad \text{or} \quad \vec{v}_{C/D} = 10\frac{2}{3} \text{ m/s} \uparrow .$$

Example 2.5

Collar A travels upward from rest with a constant acceleration, as shown in Figure 2.6. Note that no friction is present between the collars and the vertical columns. The cable is inextensible. There is no friction between the cable and pulleys. Knowing that after 6 s the relative velocity of collar B respect to collar A is 20 m/s, determine

(a) the accelerations of blocks A and B, and

(b) the velocity of block B after 6 s.

Solution:

Let y be positive downward. With reference to Figure 2.6,

$$2y_A + y_B + (y_B - y_A) = constant.$$

Then, the velocity equation, by taking the time derivative of this position equation, becomes

$$v_A + 2v_B = 0, \tag{2.13a}$$

taking the time derivative again, one obtains

$$a_A + 2a_B = 0. \tag{2.13b}$$

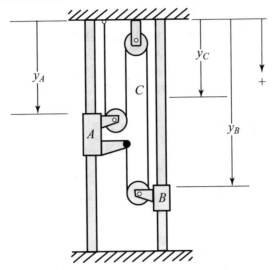

Figure 2.6: Collars traveling vertically.

(a) Applying Equation (2.13a) for the initial condition when collar A starts from rest such that

$$(v_A)_o + 2(v_B)_o = 0 \quad \text{or} \quad (v_B)_o = 0.$$

Since collar A moves upward therefore it is constant and negative. This leads, by applying Equation (2.13b), to a_B being positive or moving downward. Thus, upon integrating w.r.t. t of the acceleration term individually,

$$v_A = (v_A)_o + a_A t \quad \Rightarrow \quad v_A = a_A t$$

and

$$v_B = (v_B)_o + a_B t \quad \Rightarrow \quad v_B = a_B t$$

because $(v_A)_o = 0$, $(v_B)_o = 0$ as the system starts from rest. But, the relative velocity,

$$v_{B/A} = v_B - v_A = (a_B - a_A)t.$$

From Equation (2.13b), $a_B = -\frac{a_A}{2}$ so that

$$v_{B/A} = -\frac{3}{2}a_A t \quad \Rightarrow \quad 20 = -\frac{3}{2}a_A(6) \quad \Rightarrow \quad a_A = -\frac{20}{9} \text{ m/s}^2$$

and

$$a_B = \frac{10}{9} \text{ m/s}^2.$$

That is,

$$\vec{a}_A = \frac{20}{9} \text{ m/s}^2 \uparrow \quad \text{and} \quad \vec{a}_B = \frac{10}{9} \text{ m/s}^2 \downarrow.$$

(b) The velocity

$$v_B = a_B t = \frac{10}{9}(6) \text{ m/s} = \frac{20}{3} \text{ m/s}$$

or

$$\vec{v}_B = \frac{20}{3} \text{ m/s} \downarrow .$$

Example 2.6

A short range missile is fired from the edge of 200 m cliff with an initial velocity of 100 m/s at an angle of 35° with the horizontal, as shown in Figure 2.7a. For simplicity, air resistance is disregarded. Determine

(a) the horizontal distance x when the missile hits the ground, and

(b) the highest elevation above the ground achieved by the missile.

Solution:

The motion in the vertical and horizontal directions are considered separately in the following.

Vertical motion

This is the uniformly acceleration motion since the acceleration due to gravity is constant. Equations (2.8)–(2.10) will be applied.

The positive sense vertical and horizontal directions are shown in Figure 2.7b. Thus, the initial velocity along the y-direction is

$$(v_y)_o = (100 \text{ m/s}) \sin 35° = 57.3577 \text{ m/s}.$$

The acceleration is that due to gravity, thus

$$a = -9.81 \text{ m/s}^2.$$

Substituting into Equations (2.8)–(2.10), one has

$$v_y = (v_y)_o + at \quad \Rightarrow \quad v_y = 57.3577 - 9.81t \tag{2.14a}$$

$$y = (v_y)_o t + \frac{1}{2}at^2 \quad \Rightarrow \quad y = 57.3577t - 4.91t^2 \tag{2.14b}$$

$$v_y^2 = (v_y)_o^2 + 2ay \quad \Rightarrow \quad v_y^2 = 57.3577^2 - 19.62y. \tag{2.14c}$$

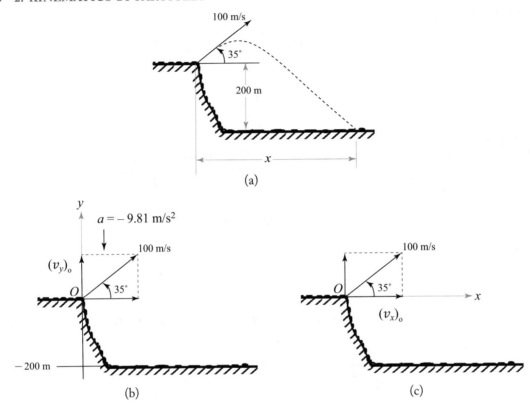

Figure 2.7: Firing of a short range missile: (a) sketch of firing, (b) vertical motion, and (c) horizontal motion.

Horizontal motion

This is the uniform motion. Thus, Equation (2.7) can be applied. With reference to Figure 2.7c, one has

$$(v_x)_o = (100 \text{ m/s}) \cos 35° = 81.9152 \text{ m/s}.$$

Therefore,

$$x = (v_x)_o \, t = 81.9152t \text{ m.} \tag{2.14d}$$

(a) When the missile hits the ground, $y = -200$ m which can be substituted into Equation (2.14b) to give

$$-200 = 57.3577t - 4.91t^2 \quad \text{or} \quad t^2 - 11.6818t - 40.7332 = 0.$$

This gives $t = 14.4925$ s or $t = -2.8107$ s. Thus, $t = 14.4925$ s is chosen and substituting this value into Equation (2.14d) to give the horizontal distance

$$x = 1187.1560 \text{ m}.$$

(b) When the missile reaches its highest elevation the vertical velocity is zero. That is,

$$0 = 57.3577^2 - 19.62y \quad \Rightarrow \quad y = \frac{57.3577^2}{19.62} \text{ m} = 167.6812 \text{ m}.$$

Therefore, the highest elevation achieved above the ground is 367.6812 m.

2.4   CURVILINEAR MOTION OF PARTICLES

Aside from rectilinear motion, a particle can undergo *curvilinear motion* in the physical world. The curvilinear motion of a particle is the motion of that particle along a curved path or trajectory. With reference to Figure 2.8, the particle P has a position vector $\vec{r}$ and the velocity $\vec{v}$ of the particle is defined as

$$\vec{v} = \frac{d\vec{r}}{dt}. \tag{2.15}$$

That is, the velocity is a *vector tangent to the path of the particle* and has a magnitude v (also called the speed) which is the time derivative of the length s of the arc described by the particle

$$v = \lim_{\Delta t \to 0} \frac{\Delta s}{\Delta t} = \frac{ds}{dt}. \tag{2.16}$$

The acceleration $\vec{a}$ of a particle is thus defined as

$$\vec{a} = \frac{d\vec{v}}{dt}. \tag{2.17}$$

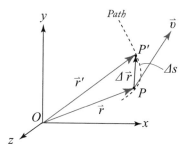

Figure 2.8: Position and velocity vectors of a particle in a curved path.

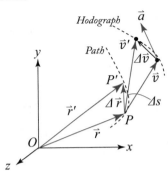

Figure 2.9: Illustration of acceleration of a particle.

In general, *the acceleration of a particle in a curved path is not a vector tangent to the path of the particle.* This point is illustrated in Figure 2.9. Note that the curve traced by the tip of velocity $\vec{v}$ is known as the *hodograph* of the motion.

For two particles A and B travelling in space as shown in Figure 2.10 the *relative position vector* of B w.r.t. A, $\vec{r}_{B/A}$, may be written as

$$\vec{r}_B = \vec{r}_A + \vec{r}_{B/A}. \tag{2.18}$$

Similarly, the *relative velocity* of B w.r.t. A, $\vec{v}_{B/A}$, may be written as

$$\vec{v}_B = \vec{v}_A + \vec{v}_{B/A} \tag{2.19}$$

and the *relative acceleration* of B w.r.t. A, $\vec{a}_{B/A}$, may be written as

$$\vec{a}_B = \vec{a}_A + \vec{a}_{B/A}. \tag{2.20}$$

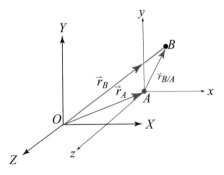

Figure 2.10: Relative motion of two particles A and B.

2.5   MOTION IN RECTANGULAR COMPONENTS

In many situations, for example in the study of motion of projectiles in space, it is convenient to express the velocity and acceleration of a particle in rectangular coordinates. Writing the rectangular coordinates of a particle P as x, y, and z, and taking the first and second derivatives w.r.t. t, one has, respectively the rectangular components of the velocity and acceleration

$$v_x = \dot{x}, \quad v_y = \dot{y}, \quad v_z = \dot{z}, \tag{2.21}$$

and

$$a_x = \ddot{x}, \quad a_y = \ddot{y}, \quad a_z = \ddot{z} \tag{2.22}$$

in which the over dot and double over dots designate, respectively, the first and second derivatives w.r.t. time t.

2.6   MOTION IN TANGENTIAL AND NORMAL COMPONENTS

In other situations, it is convenient to express the velocity and acceleration of a particle in two-dimensional (2D) space in components along the tangent to the path and the normal to the path directed toward the center of curvature of the path, as shown in Figure 2.11.

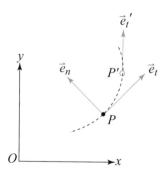

Figure 2.11: Tangential and normal components of motion a particle.

With reference to Figure 2.11, the velocity and acceleration of the particle P are expressed in terms of tangential and normal components as

$$\vec{v} = v\,\vec{e}_t, \tag{2.23}$$

where $\vec{e}_t$ is the unit vector along the tangent to the path of the particle,

$$\vec{a} = \frac{dv}{dt}\vec{e}_t + \frac{v^2}{\rho}\vec{e}_n, \tag{2.24}$$

where ρ is the radius of curvature of its path and $\vec{e}_n$ is the unit vector normal to the path of the particle.

Before the derivation of Equation (2.24) from (2.23), one needs to show that

$$\frac{d\vec{e}_t}{d\theta} = \vec{e}_n. \tag{2.25}$$

To this end, consider the unit vectors in Figure 2.12 in which the particle P moves along the curve to a new position P' in an infinitesimally small angular displacement $\Delta\theta$. Let $\vec{e}_t$ and $\vec{e}_t'$ be the two unit vectors (viewed as the radii of a circle) tangent at P and P', respectively. As $\Delta\theta$ is infinitesimally small the magnitudes of $\vec{e}_t$ and $\vec{e}_t'$ can be considered equal such that the triangle shown in Figure 2.12 is an isosceles triangle.

With reference to the latter figure,

$$\vec{e}_t' = \vec{e}_t + \Delta\vec{e}_t \quad \text{or} \quad \Delta\vec{e}_t = \vec{e}_t' - \vec{e}_t. \tag{2.26}$$

From the isosceles triangle, the magnitude of $\Delta\vec{e}_t$ becomes

$$|\Delta\vec{e}_t| = 2\sin\left(\frac{\Delta\theta}{2}\right)$$

such that upon dividing both sides by $\Delta\theta$, one has

$$\lim_{\Delta\theta\to 0}\frac{|\Delta\vec{e}_t|}{\Delta\theta} = \lim_{\Delta\theta\to 0}\frac{2\sin\left(\frac{\Delta\theta}{2}\right)}{\Delta\theta} = \lim_{\Delta\theta\to 0}\frac{\sin\left(\frac{\Delta\theta}{2}\right)}{\Delta\theta/2} = 1. \tag{2.27}$$

The vector $\Delta\vec{e}_t/\Delta\theta$ has the magnitude of 1 and direction perpendicular to the radius of the

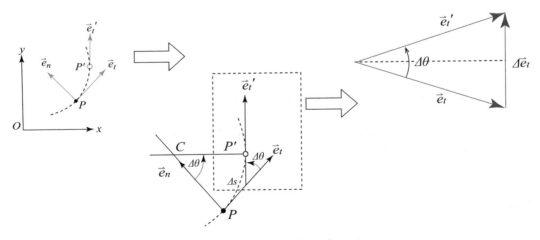

Figure 2.12: Infinitesimally small angular displacement of a unit vector.

circle or $\vec{e}_t$. Let the unit vector that is perpendicular to $\vec{e}_t$ be $\vec{e}_n$, one has

$$\vec{e}_n = \lim_{\Delta\theta\to 0} \frac{\Delta\vec{e}_t}{\Delta\theta} = \frac{d\vec{e}_t}{d\theta}.$$

This is Equation (2.25).

Now returning to the derivation of Equation (2.24), one has

$$\vec{a} = \frac{d\vec{v}}{dt} = \frac{d\left(v\vec{e}_t\right)}{dt} = \frac{dv}{dt}\vec{e}_t + v\frac{d\vec{e}_t}{dt}. \qquad (2.28)$$

Consider the second term on the rhs of Equation (2.28). Since

$$\frac{d\vec{e}_t}{dt} = \frac{d\vec{e}_t}{d\theta}\frac{d\theta}{ds}\frac{ds}{dt}$$

and from Equations (2.16) and (2.25), one obtains

$$\frac{d\vec{e}_t}{dt} = \vec{e}_n\frac{d\theta}{ds}v. \qquad (2.29)$$

With reference to Figure 2.12, the infinitesimally small arc length of the circle is

$$\Delta s = \rho\Delta\theta \quad \text{or} \quad \lim_{\Delta\theta\to 0}\frac{\Delta\theta}{\Delta s} = \frac{d\theta}{ds} = \frac{1}{\rho},$$

in which $\rho = CP = CP'$ is the radius of curvature.

Substituting this equation into Equation (2.29), and, in turn, into Equation (2.28), one arrives at

$$\vec{a} = \frac{dv}{dt}\vec{e}_t + \frac{v^2}{\rho}\vec{e}_n.$$

This is Equation (2.24). Note that the normal unit vector $\vec{e}_n$ of the particle P is always pointing toward the center of curvature of the path of the particle.

2.7   MOTION IN RADIAL AND TRANSVERSE COMPONENTS

In some problems of 2D motion, the position of the particle P is defined by its polar coordinates r and θ, as shown in Figure 2.13. In such problems it is convenient to expressed the velocity and acceleration of the particle into the components parallel and perpendicular to the line OP. These components are known as *radial* and *transverse components*.

Before expressing the velocity and acceleration into the radial and transverse components, one refers to Figure 2.13 and makes use of similar reasoning in the derivation of Equation (2.25) to give

$$\vec{e}_\theta = \frac{d\vec{e}_r}{d\theta}, \qquad \frac{d\vec{e}_\theta}{d\theta} = -\vec{e}_r. \qquad (2.30)$$

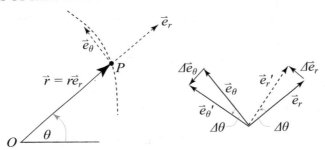

Figure 2.13: Polar coordinates of a particle in 2D motion.

The negative sign in the last equation indicates that it is opposite to the direction of $\vec{e}_r$. Thus, by using the chain rule of differentiation, one has

$$\frac{d\vec{e}_r}{dt} = \frac{d\vec{e}_r}{d\theta}\frac{d\theta}{dt} = \vec{e}_\theta \frac{d\theta}{dt}, \quad \frac{d\vec{e}_\theta}{dt} = \frac{d\vec{e}_\theta}{d\theta}\frac{d\theta}{dt} = -\vec{e}_r\frac{d\theta}{dt}$$

or applying the over dots to denote differentiation w.r.t. t

$$\dot{\vec{e}}_r = \dot{\theta}\vec{e}_\theta, \quad \dot{\vec{e}}_\theta = -\dot{\theta}\vec{e}_r. \tag{2.31}$$

Now, return to the expressions for the velocity and acceleration into their radial and transverse components. For the velocity,

$$\vec{v} = \frac{d(r\vec{e}_r)}{dt} = \dot{r}\vec{e}_r + r\dot{\vec{e}}_r = \dot{r}\vec{e}_r + r\dot{\theta}\vec{e}_\theta \quad \text{or} \quad \vec{v} = v_r\vec{e}_r + v_\theta\vec{e}_\theta. \tag{2.32}$$

For the acceleration,

$$\vec{a} = \frac{d\vec{v}}{dt} = \frac{d(\dot{r}\vec{e}_r + r\dot{\theta}\vec{e}_\theta)}{dt} = \ddot{r}\vec{e}_r + \dot{r}\dot{\vec{e}}_r + r\ddot{\theta}\vec{e}_\theta + \dot{r}\dot{\theta}\vec{e}_\theta + r\dot{\theta}\dot{\vec{e}}_\theta.$$

Substituting for Equation (2.31) and simplifying, one obtains

$$\vec{a} = \left(\ddot{r} - r\dot{\theta}^2\right)\vec{e}_r + \left(r\ddot{\theta} + 2\dot{r}\dot{\theta}\right)\vec{e}_\theta \quad \text{or} \quad \vec{a} = a_r\vec{e}_r + a_\theta\vec{e}_\theta. \tag{2.33}$$

It is important to note that a_r is not equal to the time derivative of v_r and that a_θ is not the time derivative of v_θ.

The foregoing results of velocity and acceleration can be easily extended to motion of a particle in 3D space. One approach of the motion of a particle in 3D space is by expressing the position of the particle in cylindrical coordinates R, θ, and z as illustrated in Figure 2.14 such that the position vector of particle P

$$\vec{r} = R\vec{e}_R + z\vec{k}, \tag{2.34}$$

where $\vec{k}$ is the unit vector along the axial direction and is constant in magnitude as well as direction.

Operating on Equation (2.34), one can show that

$$\vec{v} = \frac{d\vec{r}}{dt} = \dot{R}\vec{e}_R + R\dot{\theta}\vec{e}_\theta + \dot{z}\vec{k} \tag{2.35}$$

and

$$\vec{a} = \frac{d\vec{v}}{dt} = \left(\ddot{R} - R\dot{\theta}^2\right)\vec{e}_R + \left(R\ddot{\theta} + 2\dot{R}\dot{\theta}\right)\vec{e}_\theta + \ddot{z}\vec{k}. \tag{2.36}$$

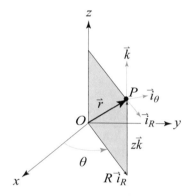

Figure 2.14: Particle motion in 3D space.

Example 2.7

A train is traveling on a curved section of track of radius 1,500 m at the speed of 200 km/h, as shown in Figure 2.15a. The brakes are suddenly applied, causing it to slow down at a constant rate. After 6 s the speed of the train is reduced to 100 km/h. Determine the acceleration of the train immediately after the brakes have been applied.

Solution:

In the present solution the tangential and normal components of the acceleration are separately considered.

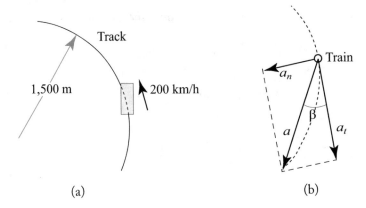

Figure 2.15: (a) Train traveling on a curved section of track and (b) direction of acceleration.

Tangential component of acceleration
Given,

$$200\frac{\text{km}}{\text{h}} = 200\frac{\text{km}}{\text{h}}\left(\frac{1,000 \text{ m}}{1 \text{ km}}\right)\left(\frac{1 \text{ h}}{3,600 \text{ s}}\right) = 55.56 \text{ m/s.}$$

Therefore,

$$100\frac{\text{km}}{\text{h}} = 27.78 \text{ m/s.}$$

Since the train slows down at a constant rate, therefore,

$$a_t = \text{average } a_t = \frac{\Delta v}{\Delta t} = \frac{(27.78 - 55.56) \text{ m/s}}{6 \text{ s}} = -4.63 \text{ m/s}^2.$$

Normal component of acceleration
The normal component of the acceleration is

$$a_n = \frac{v^2}{\rho} = \frac{55.56^2}{1500} \text{ m/s}^2 = 2.0579 \text{ m/s}^2.$$

Magnitude and direction of acceleration
The magnitude of acceleration is

$$a = \sqrt{a_t^2 + a_n^2} = \sqrt{(-4.63)^2 + (2.0579)^2}\frac{\text{m}}{\text{s}^2} = 5.067\frac{\text{m}}{\text{s}^2}.$$

With reference to Figure 2.15b, the direction is given by

$$\tan\beta = \frac{a_n}{a_t} = \frac{2.0579}{4.63} = 0.4445 \quad \Rightarrow \quad \beta = 23.96°.$$

2.8 EXERCISES

2.1. The motion of ball is governed by the relation $x = t^2 - 15t + 20$, in which the distance x is measured in m and t in s. Find (a) when the velocity is zero, and (b) the distance traveled when $t = 6$ s.

2.2. The acceleration of an oscillating particle is defined by the equation $a = -kx$. When $x = 0$ and $v = 8$ m/s, find k.

2.3. In an experiment it was found that the acceleration of a particle is governed by the relation $a = g(1 - kv^2)$. Knowing that the particle starts at $t = 0$, $x = 0$, and $v = 0$, find an equation of velocity v for any distance x.

2.4. A driver of a car traveling at 60 km/h when he observes a traffic light 200 m ahead turning red. The traffic light has been programmed to stay red for 20 s. If the driver wants to pass the light without stopping just as it turns green again, determine the required uniform deceleration of the car. It is assumed that the motion of the car is rectilinear.

2.5. Slider block A travels to the right, in Figure 2.16, with a constant velocity of 10 m/s and is connected by an inextensible cable to block B. It is assumed that no friction is present between the block and the supporting horizontal surface, and between the cable and pulley. Determine

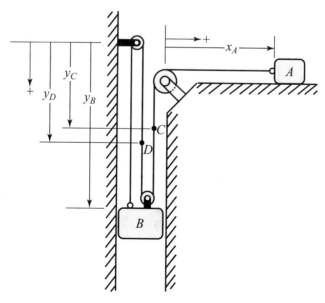

Figure 2.16: Blocks traveling horizontally and vertically.

(a) the velocity of block B,

(b) the velocity of portion D of the cable, and

(c) the relative velocity of portion C of the cable with respect to portion D.

2.6.   Slider Collar A travels to the left from rest with a constant acceleration, as shown in Figure 2.17. Note that no friction is present between the collars and the horizontal columns. The cable is inextensible. There is no friction between the cable and pulleys. Knowing that after 6 s the relative velocity of collar B respect to collar A is 10 m/s, determine

(a) the accelerations of blocks A and B, and

(b) the velocity of block B after 5 s.

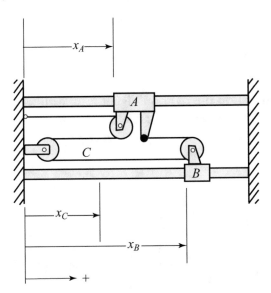

Figure 2.17: Collars traveling horizontally.

2.7.   A projectile is fired with a velocity of 200 m/s at a target A located 650 m above the gun G at a horizontal distance of 3,500 m, as shown in Figure 2.18. Disregarding air resistance, find the firing angle α.

2.8.   A fighter jet is flying horizontally, as shown in Figure 2.19, at an altitude of 3,000 m and at constant speed of 900 km/h on a trajectory which passes directly over an anti-aircraft gun which fires a shell with a muzzle velocity of 600 m/s and hits the jet. Knowing that the firing angle of the gun is 58° and disregarding air resistance, determine (a) the velocity of the shell relative to the fighter jet at the time of impact and (b) the time it takes from the gun to that at impact.

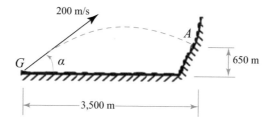

Figure 2.18: Projectile firing.

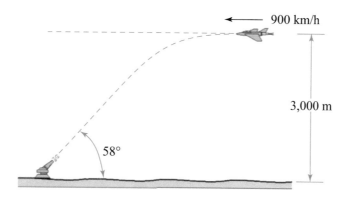

Figure 2.19: Impact of fighter jet and gun shell.

CHAPTER 3

Dynamics of Particles

3.1 INTRODUCTION

Having studied kinematics of particles in the last chapter, *dynamics* of particles is introduced in the present chapter. Briefly, dynamics is a branch of mechanics that deals with particles and bodies in motion. Specifically, Newton's laws of motion are presented in Section 3.2 while equations of motion and dynamic equilibrium of a particle are included in Section 3.3. Equations of motion in radial and transverse components are presented in Section 3.4. Momentum and rate of change of momentum of a particle are dealt with in Section 3.5. Angular momentum and rate of change of angular momentum of a particle are studied in Section 3.6. Differential equation for trajectory of a particle under a central force is included in Section 3.7.

3.2 NEWTON'S LAWS OF MOTION

Newton's laws of motion governing the motion of a particle may generally be stated as in the following three laws.

> **First Law:** *A particle continues in its state of rest or uniform motion in a straight line, unless impressed forces act on it.*
> **Second Law:** *A particle acts on by an impressed force (which is proportional to the time rate of change of linear momentum of the particle) experiences an acceleration which has the same direction as the force and a magnitude that is directly proportional to the force.*
> **Third Law:** *Action and reaction in a particle are always equal and opposite.*

Aside from the three laws stated in the foregoing, *Newton's law of gravitational attraction* is stated as follows.

> *Two given particles of matter attract each other with a force F proportional to their masses m_1, m_2, and inversely proportional to the square of distance r between them.* Mathematically,

$$F \propto \frac{m_1 m_2}{r^2} \qquad \text{or} \qquad F = G\frac{m_1 m_2}{r^2}, \tag{3.1}$$

where G is called the universal constant of gravitation and has been experimentally determined as $G = 66.73 \times 10^{-12} \ \frac{\text{m}^3}{\text{kg s}^3}$.

In engineering dynamics Newton's second law of motion is frequently applied. Symbolically, it is expressed as

$$\vec{F} = m\vec{a} \tag{3.2}$$

in which m is the mass of the particle and $\vec{a}$ is its acceleration in the direction of $\vec{F}$.

Equation (3.2) can be extended to a particle acted on simultaneously by several forces. Thus,

$$\sum \vec{F} = m\vec{a}, \tag{3.3}$$

where $\sum \vec{F}$ denotes the vector sum or resultant of all the forces acting on the particle.

3.3   EQUATIONS OF MOTION AND DYNAMIC EQUILIBRIUM

Equation (3.3) is a vector equation and for convenience it can easily be replaced by scalar equations in many problems. In the case of resolving forces into rectangular components, one writes Equation (3.3) as

$$\sum \left(F_x \vec{\imath} + F_y \vec{\jmath} + F_z \vec{k} \right) = m \left(a_x \vec{\imath} + a_y \vec{\jmath} + a_z \vec{k} \right). \tag{3.4}$$

From this vector equation it follows that three scalar equations of motion may be expressed as

$$\sum F_x = ma_x, \qquad \sum F_y = ma_y, \qquad \sum F_z = ma_z. \tag{3.5}$$

Equation (3.3) may also be written as

$$\sum \vec{F} - m\vec{a} = 0. \tag{3.6}$$

The second term on the lhs of this equation has the magnitude ma and of direction opposite to that of the acceleration, is known as the *inertia vector*. In this case the particle may be considered to be in equilibrium under the given forces and inertia vector. Thus, the particle is said to be in *dynamic equilibrium*, and the problem in question can be solved by the methods of statics (this is, loosely speaking the D'Alembert principle).

3.4   EQUATIONS OF MOTION IN RADIAL AND TRANSVERSE COMPONENTS

In addition to the equations of motion in rectangular components introduced in the last section, one can easily expressed the equations of motion in tangential and normal components, and in radial and transverse components. For conciseness, in this section only equations of motion in radial and transverse components are introduced.

Consider a particle P whose position in the 2D space is expressed in polar coordinates r and θ. With reference to the acceleration in radial and transverse components obtained in Chapter 2 and Equation (3.4), one can show that the scalar equations of motion in radial and transverse components

$$\sum F_r = ma_r, \qquad \sum F_\theta = ma_\theta. \tag{3.7}$$

Substituting for a_r and a_θ from Equation (2.33), one obtains

$$\sum F_r = m\left(\ddot{r} - r\dot{\theta}^2\right), \qquad \sum F_\theta = m\left(r\ddot{\theta} + 2\dot{r}\dot{\theta}\right). \tag{3.8}$$

Example 3.1

In a sport competition of hammer throwing the thrower swings the head H of the hammer of mass $m = 8$ kg in a horizontal circle with a constant speed v, as shown in Figure 3.1a. Given that the radius of the horizontal circle $r = 1.0$ m, the angle between the horizontal circle and the rope HO, $\theta = 45°$, determine the tension in the rope HO.

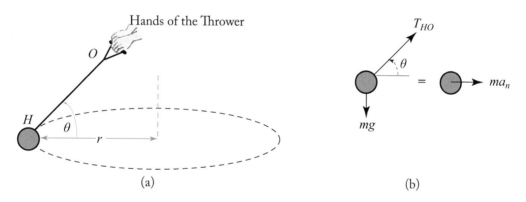

Figure 3.1: (a) Thrower swings hammer mass and (b) FBD.

Solution:
Given data are $m = 8$ kg, $r = 1.0$ m, $\theta = 45°$, and v is constant. The free-body diagram (FBD) is included in Figure 3.1b.
Let the tension in the rope be T_{HO}.

With reference to Figure 3.1b, the acceleration at H is given by

$$\vec{a} = \vec{a}_n + \vec{a}_t = \vec{a}_n, \qquad \vec{a}_t = 0 \text{ because } v \text{ is constant.}$$

Therefore, the magnitude of $\vec{a}$

$$a = a_n = \frac{v^2}{r}.$$

$+\uparrow \quad \sum F_y = 0,$

$$T_{HO} \sin 45° - mg = 0.$$

This equation gives the tension in the rope,

$$T_{HO} = \frac{mg}{\sin 45°} = \frac{(8 \text{ kg})(9.81 \text{ m/s}^2)}{0.7071} = 110.99 \text{ N}.$$

Example 3.2
A number of packages, each of mass $m = 1.0$ kg, is being moved at a velocity $v = 1.0$ m/s by the conveyor belt, shown in Figure 3.2a. If the coefficient of static friction between each package and the conveyor belt is $\mu = 0.5$, determine the force acts by the belt on the package just after the package passed point A.

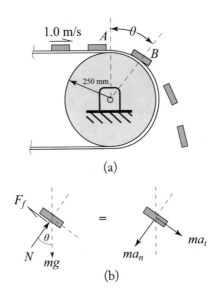

Figure 3.2: (a) Packages on a moving conveyor belt and (b) FBD of package at B.

Solution:
Assume packages *do not slip* so that the tangential component of the acceleration $\vec{a}_t = 0$, and the frictional force satisfies $F_f \leq \mu N$.

With reference to the FBD in Figure 3.2b, the acceleration on the curved portion of the belt is

$$a = a_n = \frac{v^2}{r} = \frac{(1 \text{ m/s})^2}{0.250 \text{ m}} = 4.0 \text{ m/s}$$

and resolving forces $+ \nearrow \quad \sum F_y = ma_y$, one has

$$N - mg \cos \theta = -ma_n$$

in which the negative sign on the rhs indicates that the normal component of the acceleration is directed toward the center of curvature. Therefore,

$$N = mg \cos \theta - m \left(\frac{v^2}{r} \right), \qquad (3.9a)$$

$+ \searrow \quad \sum F_x = ma_x$, one obtains

$$- F_f + mg \sin \theta = ma_t = 0. \qquad (3.9b)$$

Note that in Equation (3.9b) the negative sign in front of the frictional force implies that it is opposite to the positive x-axis in Figure 3.2b while the rhs is zero because the tangential component of the acceleration $\vec{a}_t = 0$.

Applying Equations (3.9a), (3.9b), and the condition that $F_f = \mu N$, one obtains

$$mg \cos \theta - m \left(\frac{v^2}{r} \right) = \frac{mg \sin \theta}{\mu} \qquad \text{or} \qquad g \cos \theta - \frac{v^2}{r} = \frac{g \sin \theta}{\mu}. \qquad (3.9c)$$

At point A, $\theta = 0$. Therefore, from Equation (3.9a), one has

$$N = mg - m \left(\frac{v^2}{r} \right) = (1 \text{ kg}) \left(9.81 \text{ m/s}^2 \right) - (1 \text{ kg}) \frac{(1.0 \text{ m/s})^2}{0.250 \text{ m}}$$

$$= 5.81 \text{ N}.$$

This is the force acted by the belt on the package.

Example 3.3
The cable system shown in Figure 3.3a is at rest when a constant force $P = 300$ N is applied to block A. Disregarding the masses of the pulleys, effect of friction in the pulleys, and assuming that the *static* and *kinetic coefficients* of friction between block A and the horizontal surface are respectively, $\mu_s = 0.30$ and $\mu_k = 0.25$, determine the friction force F_{AK} when block A is sliding. Given that $m_A = 30$ kg, $m_B = 25$ kg.

Solution:
Let F be the tension in the cable. First, it is necessary to check the equilibrium position to see if the blocks move.

For block B, with reference to the FBD in Figure 3.3b and recall that the system is initially at rest,

$$3F - m_B g = 0 \quad \Rightarrow \quad F = \frac{(25 \text{ kg}) \left(9.81 \text{ m/s}^2 \right)}{3} = 81.75 \text{ N}.$$

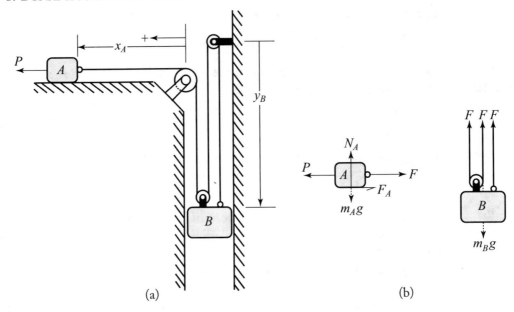

Figure 3.3: (a) Cable system and (b) FBD of cable system.

For block A, $+ \uparrow$ $\sum F_y = 0$,

$$N_A - m_A g = 0 \quad \Rightarrow \quad N_A = m_A g = (30 \text{ kg}) \left(9.81 \text{ m/s}^2\right) = 294.3 \text{ N}.$$

For block A, $+ \leftarrow$ $\sum F_x = 0$,

$$300 - F_A - F = 0 \quad \Rightarrow \quad F_A = 300 - 81.75 \text{ N} = 168.25 \text{ N}.$$

The force due to the static friction is

$$\mu_s \, N_A = (0.30)(294.3) \text{ N} = 88.29 \text{ N}.$$

This means that $F_A > \mu_s \, N_A$ and therefore the blocks move.
Thus, the friction force during sliding is

$$F_{AK} = \mu_k \, N_A = (0.25)(294.3 \text{ N}) = 73.575 \text{ N}.$$

Example 3.4
A curve in a speed test track has a radius of 300 m and a rated speed of 190 km/h (note that the rated speed of banked highway curve is that at which a car must travel when there is no lateral friction acting on the wheels of the car). If a car running at 290 km/h in this test track starts skidding on the curve, determine

(a) the banking angle θ, as shown in Figure 3.4a;

(b) the coefficient of static friction μ_s between the tires of the car and the track surface under the prevailing conditions (that is, at the banking angle found and speed of 290 km/h); and

(c) the minimum speed at which the same car could just manage to move on the curved track.

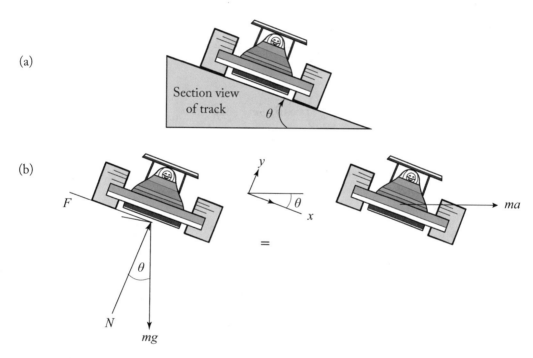

Figure 3.4: (a) Banking angle of a curve test track and (b) FBD of car on track.

Solution:
Let the coordinate system be shown in Figure 3.4b. The weight of the car

$$W = mg,$$

and acceleration

$$a = \frac{v^2}{r}.$$

$$+\searrow \quad \sum F_x = ma_x, \qquad\qquad F + W \sin\theta = ma\cos\theta$$

$$\Rightarrow \quad F = m\left(\frac{v^2}{r}\right)\cos\theta - mg\sin\theta \qquad (3.10a)$$

$$+\nearrow \quad \sum F_y = ma_y, \qquad\qquad N - W\cos\theta = ma\sin\theta$$

$$\Rightarrow \quad N = m\left(\frac{v^2}{r}\right)\sin\theta + mg\cos\theta. \qquad (3.10b)$$

(a) *Banking angle*

Rated speed $v = 190\ \frac{\text{km}}{\text{h}} = \frac{(190)(1{,}000\ \text{m})}{3{,}600\ \text{s}} = 52.78\ \frac{\text{m}}{\text{s}}$. At rated speed, $F = 0$.

Substituting into Equation (3.10a), one has

$$0 = m\left(\frac{v^2}{r}\right)\cos\theta - mg\sin\theta \quad \Rightarrow \quad \tan\theta = \frac{v^2}{rg}$$

$$\Rightarrow \quad \tan\theta = \frac{\left(52.78\ \frac{\text{m}}{\text{s}}\right)^2}{(300\ \text{m})\left(9.81\ \text{m/s}^2\right)} = 0.9466 \quad \Rightarrow \quad \theta = 43.43°.$$

(b) *Coefficient of static friction*

Under the prevailing conditions, $v = 290\frac{\text{km}}{\text{h}} = \frac{(290)(1{,}000\ \text{m})}{3{,}600\ \text{s}} = 80.56\ \frac{\text{m}}{\text{s}}$.

For $F = \mu_s N$, by making use of Equations (3.10a) and (3.10b), one obtains

$$\mu_s = \frac{F}{N} = \frac{m\left(\frac{v^2}{r}\right)\cos\theta - mg\sin\theta}{m\left(\frac{v^2}{r}\right)\sin\theta + mg\cos\theta} = \frac{v^2\cos\theta - rg\sin\theta}{v^2\sin\theta + rg\cos\theta}$$

$$\Rightarrow \quad \mu_s = \frac{(80.56^2)\cos 43.43° - (300)(9.81)\sin 43.43°}{(80.56^2)\sin 43.43° + (300)(9.81)\cos 43.43°}$$

$$\Rightarrow \quad \mu_s = \frac{(6489.9136)(0.7262) - (2943)(0.6875)}{(6489.9136)(0.6875) + (2943)(0.7262)}$$

$$\Rightarrow \quad \mu_s = 0.4076 \quad \text{which is the required answer.}$$

(c) *Minimum speed*

This requires that $F = -\mu_s N$ such that $\mu_s = \frac{-F}{N} = \frac{-v^2\cos\theta + rg\sin\theta}{v^2\sin\theta + rg\cos\theta}$.

This gives

$$v^2 = \frac{rg\,(\sin\theta - \mu_s\cos\theta)}{\mu_s\sin\theta + \cos\theta} = \frac{(300)(9.81)[0.6875 - (0.4076)(0.7262)]}{(0.4076)(0.6875) + 0.7262}$$

$$\Rightarrow \quad v^2 = \frac{(2943)\,[0.6875 - (0.4076)\,(0.7262)]}{(0.4076)\,(0.6875) + 0.7262} = \frac{1152.1871}{1.0064}$$

$$\Rightarrow \quad v = 33.8358 \text{ m/s.}$$

3.5 MOMENTUM AND RATE OF CHANGE OF MOMENTUM OF A PARTICLE

In Equation (3.3), one can replace the acceleration $\vec{a}$ with the time derivative of velocity such that it becomes

$$\sum\vec{F} = m\vec{a} = m\frac{d\vec{v}}{dt}$$

or, because the mass m of the particle is constant,

$$\sum\vec{F} = \frac{d(m\vec{v})}{dt} \qquad (3.11)$$

in which $m\vec{v}$ is called the *linear momentum* or simply *momentum* of the particle and its direction is the same as its velocity.

Equation (3.11) states that the *resultant of the forces acting on the particle is equal to the rate of change of linear momentum of the particle*. In fact, this is the original form of Newton's second law of motion. In many textbooks of engineering dynamics the linear momentum of the particle is represented by

$$\vec{L} = m\vec{v} \qquad (3.12)$$

and therefore the resultant of the forces acting on the particle is written as

$$\sum\vec{F} = \frac{d\vec{L}}{dt} = \dot{\vec{L}}. \qquad (3.13)$$

This equation provides another means of expressing the resultant forces acting on the particle in term of the time derivative of linear momentum.

3.6 ANGULAR MOMENTUM AND RATE OF CHANGE OF ANGULAR MOMENTUM OF A PARTICLE

Now, the concept of angular momentum and the rate of change of angular momentum are considered in this section. Suppose the particle P of mass m moves w.r.t. a Newtonian or inertial frame of reference (*NFR*)*Oxyz*, as shown in Figure 3.5.

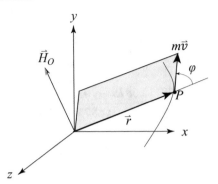

Figure 3.5: Angular momentum of a particle.

The moment about the origin O of the vector $m\vec{v}$ is called the *moment of momentum* or the *angular momentum* of the particle about O at that instant. It is represented as

$$\vec{H}_o = \vec{r} \times m\vec{v} \tag{3.14}$$

which expresses the angular momentum $\vec{H}_o$ of the particle as the vector cross-product of the position vector $\vec{r}$ and the linear momentum $m\vec{v}$. Thus, $\vec{H}_o$ is a vector perpendicular to the plane containing the position vector $\vec{r}$ and the linear momentum $m\vec{v}$. In other words, the magnitude of the angular momentum is given as

$$H_o = rmv \sin\varphi = rm\left(r\dot{\theta}\right) = mr^2\dot{\theta}, \tag{3.15}$$

where φ is the angle between the position vector $\vec{r}$ and the linear momentum $m\vec{v}$. The sense of $\vec{H}_o$ is determined by applying the right-hand rule.

In many dynamic problems, $\vec{H}_o$ can be expressed in the rectangular coordinate system as

$$\vec{H}_o = H_x\vec{i} + H_y\vec{j} + H_z\vec{k} = \vec{r} \times m\vec{v} = \begin{vmatrix} \vec{i} & \vec{j} & \vec{k} \\ x & y & z \\ mv_x & mv_y & mv_z \end{vmatrix}, \tag{3.16}$$

and operating on this determinant, one has

$$H_x = m\left(yv_z - zv_y\right), \quad H_y = m\left(zv_x - xv_z\right), \quad H_z = m\left(xv_y - yv_x\right). \tag{3.17}$$

Clearly, in the case of motion of a particle in the xy-plane so that $z = 0$, $v_z = 0$ which lead to $H_x = 0$ as well as $H_y = 0$, the angular momentum is perpendicular to the xy-plane and completely defined by the scalar quantity

$$H_o = H_z = m\left(xv_y - yv_x\right). \tag{3.18}$$

Now, consider the rate of change of angular momentum of a particle. Taking the derivative of Equation (3.16) w.r.t. t, one obtains

$$\frac{d\vec{H}_o}{dt} = \frac{d\vec{r}}{dt} \times m\vec{v} + \vec{r} \times m\frac{d\vec{v}}{dt} = \vec{v} \times m\vec{v} + \vec{r} \times m\vec{a}. \qquad (3.19)$$

The first term on the rhs is zero because $\vec{v}$ and $m\vec{v}$ are collinear. On the other hand, $m\vec{a} = \sum \vec{F}$ by Newton's second law of motion. Therefore, Equation (3.19) reduces to

$$\frac{d\vec{H}_o}{dt} = \vec{r} \times m\vec{a} = \vec{r} \times \sum \vec{F} = \sum \vec{M}_o \quad \text{or} \quad \sum \vec{M}_o = \dot{\vec{H}}_o. \qquad (3.20)$$

This equation expresses the *sum of the moments of the forces acting on the particle about the origin O as the rate of change of angular momentum of the particle about O*.

3.7 DIFFERENTIAL EQUATION FOR TRAJECTORY OF A PARTICLE UNDER A CENTRAL FORCE

The objective of this section is to derive a differential equation that predicts or provides the trajectory of the particle under a central force. Areas of application are in space mechanics in which the orbit of a satellite is of great importance.

To begin with, one considers a particle P under a central force $\vec{F}$, as shown in Figure 3.6.

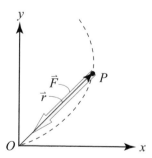

Figure 3.6: Particle under a central force.

Recall that the position vector $\vec{r}$ measures positive from the origin O of the *NFR* to the particle P. Note that the central force as indicated in the figure is in the opposite direction and therefore the central force is considered negative. Recall also that the radial and transverse components of the force of the particle in polar coordinates are given by Equation (3.8) as

$$\sum F_r = m\left(\ddot{r} - r\dot{\theta}^2\right), \qquad \sum F_\theta = m\left(r\ddot{\theta} + 2\dot{r}\dot{\theta}\right).$$

With reference to Figure 3.6,

$$\sum F_r = -F = m\left(\ddot{r} - r\dot{\theta}^2\right), \qquad \sum F_\theta = 0. \tag{3.21}$$

For the particle under a central Equation (3.20) gives

$$\dot{\vec{H}} = 0 \qquad \text{since} \qquad \sum \vec{M}_o = 0.$$

Upon integration w.r.t. t,

$$\vec{H}_o = h$$

where h is a constant and by making use of Equation (3.15)

$$\vec{H}_o = h = mr^2\dot{\theta}.$$

Rewriting this equation,

$$\dot{\theta} = \frac{d\theta}{dt} = \frac{h}{mr^2} = \frac{c}{r^2} \tag{3.22}$$

in which c is a constant since h and m are constant. Thus, c is a constant representing the angular momentum per unit mass.

By definition,

$$\dot{r} = \frac{dr}{dt} = \frac{dr}{d\theta}\frac{d\theta}{dt} = \frac{dr}{d\theta}\frac{c}{r^2}. \tag{3.23}$$

Also,

$$\frac{d\left(\frac{1}{r}\right)}{d\theta} = \frac{d\left(\frac{1}{r}\right)}{dr}\frac{dr}{d\theta} = -\frac{1}{r^2}\frac{dr}{d\theta}. \tag{3.24}$$

Substituting this equation into (3.23), one obtains

$$\dot{r} = \frac{dr}{d\theta}\frac{c}{r^2} = -r^2\frac{d\left(\frac{1}{r}\right)}{d\theta}\frac{c}{r^2} = -c\frac{d\left(\frac{1}{r}\right)}{d\theta}. \tag{3.25}$$

Taking the time derivative of $\dot{r}$ and applying Equation (3.22),

$$\ddot{r} = \frac{d\dot{r}}{dt} = \frac{d\dot{r}}{d\theta}\frac{d\theta}{dt} = \frac{d\dot{r}}{d\theta}\frac{c}{r^2}.$$

Substituting for Equation (3.25), one has

$$\ddot{r} = \frac{d\dot{r}}{d\theta}\frac{c}{r^2} = \frac{d\left[-c\frac{d(1/r)}{d\theta}\right]}{d\theta}\frac{c}{r^2} = -\left(\frac{c}{r}\right)^2\frac{d^2\left(\frac{1}{r}\right)}{d\theta^2}. \tag{3.26}$$

Now, writing $u = 1/r$ and substituting Equations (3.22) and (3.26) into the first equation of (3.21), one obtains

$$-F = m\left[-c^2u^2\frac{d^2u}{d\theta^2} - r\left(cu^2\right)^2\right]$$

or

$$-F = m\left[-c^2u^2\frac{d^2u}{d\theta^2} - \frac{1}{u}\left(cu^2\right)^2\right].$$

Simplifying, it becomes

$$\frac{d^2u}{d\theta^2} + u = \frac{F}{mc^2u^2}. \qquad (3.27)$$

This is the differential equation for the trajectory of a particle under a central force.

3.8   EXERCISES

3.1. In a spherical ball of mass $m = 10$ kg, as shown in Figure 3.7, is attached to a cable of length $\ell = 3$ m and is made to revolve in a horizontal circle at a constant velocity v_o. If the cable forms an angle $\theta = 45°$ with the horizontal circle, find (a) tension in the cable and (b) the velocity v_o of the spherical ball.

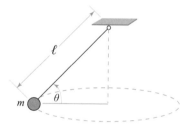

Figure 3.7: Revolving spherical ball.

3.2. The cable system shown in Figure 3.8 is at rest when a constant force $P = 300$ N is applied to block A. Disregarding the masses of the pulleys, effect of friction in the pulleys, and assuming that the *static* and *kinetic coefficients* of friction between block A and the horizontal surface are, respectively, $\mu_s = 0.30$ and $\mu_k = 0.25$, determine the friction force F_{AK} when block A is sliding. Given that $m_A = 30$ kg, $m_B = m_C = 25$ kg.

3.3. A trailer truck with a 1,000 kg cab and a 12,000 kg trailer is traveling on a level road at 100 km/h. In attempting to stop the trailer truck the driver discovers that the brakes on the trailer fail and the antiskid system of the cab gives the largest possible force that does not cause the wheels of the cab to slide. If the coefficient of static friction is 0.50, determine

(a) the shortest time for the rig to come to a stop, and

(b) the force in the coupling during such a shortest time.

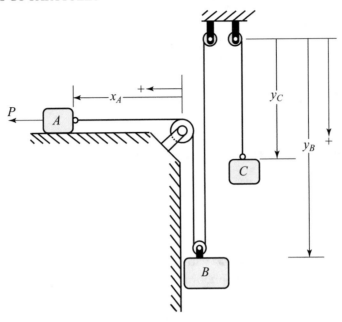

Figure 3.8: Cable system with three block masses.

3.4.   A curve in a speed test track has a radius of 500 m and a rated speed of 180 km/h (note that the rated speed of banked highway curve is that at which a car must travel when there is no lateral friction acting on the wheels of the car). If a car moving at 250 km/h in the test track starts skidding on the curve, determine

   (a) the banking angle θ, as shown in Figure 3.9;

   (b) the coefficient of static friction μ_s between the tires of the car and the track surface under the prevailing conditions (that is, at the banking angle found and speed of 250 km/h); and

   (c) the minimum speed at which the same car could just manage to move on the curved track.

3.5.   A 1.0 kg spherical ball slides on a smooth horizontal table at the free end of a cord which passes through a frictionless hole O at the center of the table, as shown in Figure 3.10. If the speed of the ball is $v_1 = 2$ m/s when the length cord on the surface of the table is $r_1 = 0.30$ m. If the known breaking strength of the cord is 100 N, find (a) the shortest length r_2 which can be reached by slowly drawing (such that the angular momentum is constant) the cord through the hole and (b) the corresponding speed v_2.

3.6.   A small ball swings in a horizontal circle at the end of the cord, as shown in Figure 3.11. The cord forms an angle θ_1 with the horizontal circle whose radius is r_1. It is drawn

Figure 3.9: Banking angle and curve test track.

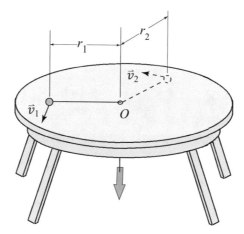

Figure 3.10: Revolving ball on a smooth table.

slowly (such that the angular momentum is constant) through a horizontal support at O. Subsequently, the cord forms a horizontal circle of radius r_2. (a) Derive a relation for r_1, r_2, θ_1, and θ_2, and (b) determine the radius r_2 for which $\theta_2 = 45°$ if the ball is set in motion initially such that $r_1 = 0.5$ m and $\theta_1 = 25°$.

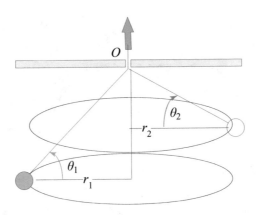

Figure 3.11: Revolving small ball being drawn slowly upward.

CHAPTER 4

Work and Energy of Particles

4.1 INTRODUCTION

In the foregoing chapters direct application of Newton's second law of motion and principles of kinematics enables one to determine the position, velocity, and acceleration of a particle in motion. In this chapter, methods for relating force, mass, displacement, and velocity are introduced.

Specifically, work of a force and potential energy of a particle are introduced in Section 4.2. Section 4.3 is concerned with potential energy and strain energy.

Kinetic energy of a particle and the principle of work and energy are presented in Section 4.4. The principle of conservation of energy is included in Section 4.5. Power and mechanical efficiency are briefly mentioned in Section 4.6.

4.2 WORK OF A FORCE AND POTENTIAL ENERGY

Consider a particle moving from a point P to a neighboring point P', as shown in Figure 4.1.

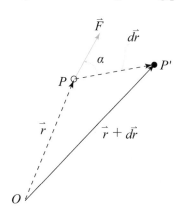

Figure 4.1: Force acting on a particle moving from one point to another.

Let $\vec{r}$ be the position vector measuring from the origin of the frame of reference to the particle at P, and $d\vec{r}$ (strictly speaking, it should be $\Delta\vec{r}$ and in the limit $\Delta\vec{r}$ approaches $d\vec{r}$) be the *displacement* vector (small vector joining P and P' in Figure 4.1) of the particle. If a force $\vec{F}$ acts on the particle the work of the force associated with the displacement $d\vec{r}$ may be defined as the dot product of two vectors,

$$dU = \vec{F} \cdot d\vec{r} \tag{4.1a}$$

which is a scalar quantity and this equation can be written as

$$dU = F \, ds \cos\alpha, \tag{4.1b}$$

where F, ds, and α are, respectively, the magnitudes of the force, displacement, and angle enclosing by $\vec{F}$ and $d\vec{r}$.

Equations (4.1a) and (4.1b) can be generalized to the case shown in Figure 4.2 in which the particle at position P_1 moves to position P_2 in a *finite* displacement. The work in this finite displacement is denoted by $U_{1\rightarrow 2}$,

$$U_{1\rightarrow 2} = \int_{P_1}^{P_2} \vec{F} \cdot d\vec{r} \tag{4.2a}$$

which can be rewritten as

$$U_{1\rightarrow 2} = \int_{s_1}^{s_2} F \cos\alpha \, ds = \int_{s_1}^{s_2} F_t \, ds \tag{4.2b}$$

by using Equation (4.1b) since $F \cos\alpha$ represents the tangential component F_t of the force. Note that in this equation the integration is performed along the path traveled by the particle.

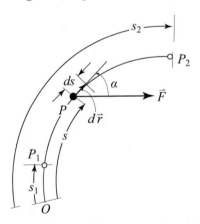

Figure 4.2: Force acts on a particle that moves in a finite displacement.

The foregoing equations can be applied to the work of a particle of mass m under the influence of gravity, as shown in Figure 4.3. That is, the work of the weight $\vec{W} = m\vec{g}$ of the particle is obtained by replacing $\vec{F}$ with $\vec{W}$ in Equation (4.2a), for example.

With reference to Figure 4.3 and applying Equations (4.2a) and (4.2b), the work of the particle P is obtained as

$$U_{1\to2} = -\int_{y_1}^{y_2} W\,dy = -Wy_2 - (-Wy_1) = Wy_1 - Wy_2$$

or

$$U_{1\to2} = mg(y_1 - y_2) \qquad (4.3)$$

in which $W = mg$ has been used. Note that the negative sign in front of the integral is due to the fact that the y-axis is chosen to be positive upward in Figure 4.3. Thus, the *work of the weight is equal to the product of mg and vertical displacement of the particle*. This work is referred to as *potential* energy.

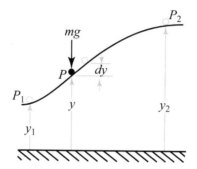

Figure 4.3: Weight acts on a particle that moves vertically.

4.3   POTENTIAL ENERGY AND STRAIN ENERGY

In Figure 4.4 a block of mass m is connected by a spring to a fixed point O. This block is supported by frictionless wheels and is moved to the right horizontally by a force $\vec{F}$. Conceptually, the block is considered as a particle. By Hooke's law the force in the deformed spring is given by

$$F = kx,$$

where k is called the *spring constant* or *spring coefficient* and x is the *deflection* of the spring. The work of the *restoring force* in the spring undergoing a finite deflection is defined by

$$U_{1\to2} = -\int_{x_1}^{x_2} F\,ds = -\int_{x_1}^{x_2} kx\,ds = \frac{1}{2}kx_1^2 - \frac{1}{2}kx_2^2, \qquad (4.4)$$

where the negative sign indicates that the restoring force in the spring is opposite to the applied force $\vec{F}$ which is positive along the increasing x coordinate in Figure 4.4. The terms on the rhs of Equation (4.4) are called the *strain energies*.

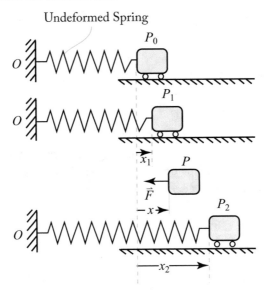

Figure 4.4: Block connected by undeformed spring undergoing extension.

4.4 KINETIC ENERGY OF A PARTICLE AND PRINCIPLE OF WORK AND ENERGY

In this section kinetic energy of a particle in translation is considered. Suppose a particle of mass m moves along a path (rectilinear or curved) and is acted on by a force $\vec{F}$, as shown in Figure 4.5.

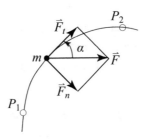

Figure 4.5: A particle moves along a path and is acted on by a force.

With reference to Figure 4.5, one may write Newton's second law in terms of the tangential component of the force as

$$F_t = m\,a_t = m\frac{dv}{dt},$$

where v is the speed of the particle and is given by Equation (2.16) as $v = \frac{ds}{dt}$. The above equation can be written as

$$F_t = m\frac{dv}{dt} = m\frac{dv}{ds}\frac{ds}{dt} = m\,v\frac{dv}{ds}.$$

Substituting this equation into Equation (4.2b) for the work of the particle, one has

$$U_{1\to 2} = \int_{s_1}^{s_2} F_t\, ds = \int_{v_1}^{v_2} m\,v\, dv = \frac{1}{2}m\,v_2^2 - \frac{1}{2}m\,v_1^2$$

or simply

$$U_{1\to 2} = T_2 - T_1 \tag{4.5}$$

in which the kinetic energy of the particle is denoted by $T_i = \frac{1}{2}mv_i^2$, $i = 1, 2$. This equation states that *the work of the force $\vec{F}$ is equal to the change in kinetic energy of the particle*. This is known as the *principle of work and energy*.

Rearranging terms in Equation (4.5), one has

$$T_1 + U_{1\to 2} = T_2. \tag{4.6}$$

This equation states that *the sum of the kinetic energy at position P_1 and the work done by the force $\vec{F}$ during the displacement from P_1 to P_2 is equal to the kinetic energy of the particle at position P_2*.

As both work and kinetic energy are scalar quantities and therefore when several forces act on the particle the expression $U_{1\to 2}$ represents the total work of the forces acting on the particle. That is, $U_{1\to 2}$ is obtained as an algebraic sum of the work of the various forces.

4.5   PRINCIPLE OF CONSERVATION OF ENERGY

In the preceding sections if the force $\vec{F}$ acting on the particle is *conservative*, meaning if the work $U_{1\to 2}$ is independent of the path traced by the particle P as it moves from position P_1 to P_2, one can write

$$U_{1\to 2} = V_1 - V_2 \tag{4.7}$$

in which V_i, $i = 1, 2$ are the potential energy of the conservative force (for example, the weight of the particle or the force exerted by a spring) acting on the particle. In this case, the principle of work and energy defined by Equation (4.5) can be modified to

$$U_{1\to 2} = V_1 - V_2 = T_2 - T_1$$

or

$$T_1 + V_1 = T_2 + V_2. \tag{4.8}$$

This is known as the *principle of conservation of energy*. It states that the *total energy (sum of kinetic energy and potential energy) of a particle acting on by a conservative force or forces is constant.*

Remarks:
The main difference between the principle of work and energy, Equation (4.6), and the principle of conservation of energy, Equation (4.8), is that Equation (4.6) applies to cases in which the forces can be *non-conservative* whereas Equation (4.8) applies to a particle acting on by a conservative force or forces.

4.6 POWER AND MECHANICAL EFFICIENCY

Power is defined as the time rate of work. If ΔU is the incremental work done during the incremental interval Δt the average power during the interval is

$$\Delta \gamma = \frac{\Delta U}{\Delta t}$$

so that power is given as

$$\gamma = \lim_{\Delta t \to 0} \frac{\Delta U}{\Delta t} = \frac{dU}{dt}.$$

Substituting for Equation (4.1a), one has

$$\gamma = \frac{dU}{dt} = \frac{\vec{F} \cdot d\vec{r}}{dt} = \vec{F} \cdot \vec{v} \tag{4.9}$$

which states the *power is the scalar or dot product of force and velocity.*

The mechanical efficiency η of a machine is defined as the ratio of output power γ_0 to the input power γ_I

$$\eta = \frac{\gamma_0}{\gamma_I}. \tag{4.10}$$

Note that since power is a scalar quantity the mechanical efficiency of a machine is also a *dimensionless* scalar quantity.

Example 4.1
The cable system shown in Figure 3.3a is at rest when a constant force $P = 300$ N is applied to block A. Disregarding the masses of the pulleys, effect of friction in the pulleys, and assuming that the *static* and *kinetic coefficients* of friction between block A and the horizontal surface are respectively, $\mu_s = 0.30$ and $\mu_k = 0.25$, determine

(a) the velocity of block B after block A has moved 3 m, and

(b) the tension in the cable.

Solution:

The analysis in the solution to Example 3.3 is applied in the present problem. With reference to Figure 3.3b, the *constraint condition* of the cable is

$$x_A + 3y_B = constant.$$

Taking the time derivative of this equation gives

$$v_A + 3v_B = 0 \quad \Rightarrow \quad v_A = -3v_B. \tag{4.11a}$$

Recall, $m_A = 30$ kg, $P = 300$ N. The principle of work and energy defined by Equation (4.6) is applicable.

For block A, $(T_1)_A + (U_{1 \rightarrow 2})_A = (T_2)_A$ in which $(T_1)_A = 0$ since the cable system starts from rest. The work done during the movement of 3 m is

$$(U_{1 \rightarrow 2})_A = (P - F_A - F)(3 \text{ m}).$$

Therefore, $(P - F_A - F)(3 \text{ m}) = \frac{1}{2}m_A(v_A^2)$.

By Equation (4.11a) and the results in Example 3.3,

$$(300 \text{ N} - 168.25 \text{ N} - F)(3 \text{ m}) = \frac{1}{2}(30 \text{ kg})(9v_B^2)$$

$$\Rightarrow \quad 395.25 - 3F = 135(v_B^2), \tag{4.11b}$$

For block B, $m_B = 25$ kg, $m_B g = 245.25$ N. Applying Equation (4.6), $(T_1)_B + (U_{1 \rightarrow 2})_B = (T_2)_B$, $(T_1)_B = 0$ since the cable system starts from rest.

The work done during the movement of 3 m is

$$(U_{1 \rightarrow 2})_B = (3F - m_B g)(-\Delta y_B)$$

$$\Rightarrow \quad (U_{1 \rightarrow 2})_B = (3F - 245.25 \text{ N}) \left(\frac{\Delta x_A}{3} \right),$$

using the constraint condition. Therefore,

$$(3F - 245.25 \text{ N}) \left(\frac{3 \text{ m}}{3} \right) = \frac{1}{2}(25 \text{ kg})(v_B^2) = 12.5(v_B^2). \tag{4.11c}$$

Adding Equations (4.11c) and (4.11b), one has

$$150 = 147.5(v_B^2) \quad \Rightarrow \quad v_B = 1.0084 \text{ m/s}.$$

(a) The velocity of block B is

$$\vec{v}_B = 1.0084 \text{ m/s} \quad \uparrow \, .$$

(b) Substituting v_B into Equation (4.11b), one obtains the tension in the cable

$$F = 85.9873 \text{ N.}$$

That is, $\vec{F} = 85.9873 \text{ N}$ ↑.

Example 4.2

As shown in Figure 4.6a, a spring device consisting of one rigid panel connected by cables on both sides to hold on the initial 50 mm compression of the spring. This spring device is used as a crash-resistant wall to prevent the important structure (not shown) further down the incline plane of 20° to the horizontal line. If a 30 kg block having a velocity $v = 2$ m/s traveling 8 m away from and toward the panel of the spring device, determine the additional deflection of the spring device in bringing the block to rest. Assume that the *kinetic coefficient of friction* between the block and the incline is 0.2 and the stiffness constant of the spring device is $k = 20$ kN/m.

Solution:

The given data are:

$k = 20,000$ N/m, $m = 30$ kg,

$v = 2$ m/s, $\theta = 20°$, initial compression of spring device,

$x_0 = 50$ mm, $\mu_k = 0.2$, and let the additional deflection of the spring device be x.

Since friction is included in the present problem and therefore, work energy principle, Equation (4.6) is applied

$$T_1 + U_{1\to 2} = T_2 \tag{4.12}$$

in which $U_{1\to 2}$ represents the work, strain energy, and energy attenuated by friction at the two stages of motion.

Kinetic energy at stage 1

$$T_1 = \frac{1}{2}mv^2 = \frac{1}{2}(30)(2^2) \text{ N.m} = 60 \text{ J.}$$

Kinetic energy at stage 2

$$T_2 = \frac{1}{2}mv_2^2 = \frac{1}{2}(30)(0^2) \text{ N.m} = 0 \text{ since the block is brought to rest.}$$

Work in two stages

$$U_{1\to 2} = U_1 - U_2 - (U_{1\to 2})_f,$$

where

$$U_1 = U_{1g} + U_{1e} = mgh_1 + \frac{1}{2}k\left(x_0^2\right),$$

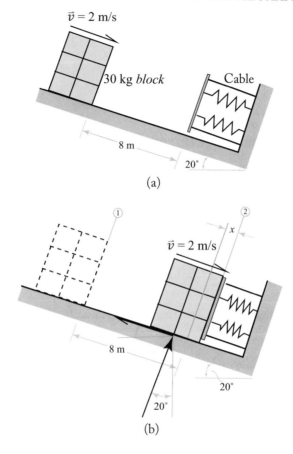

Figure 4.6: (a) Spring device and traveling block, and (b) stages of motion of traveling block.

in which

$$U_{1g} = mgh_1 = -30(9.81)\left(8\sin 20°\right) \text{ J}, \; U_{1e} = \frac{1}{2}(20{,}000)(0.05^2) \text{ J}.$$

Substituting, $U_1 = (805.254 + 25) \text{ J}.$

$$U_2 = U_{2g} + U_{2e} = mgh_2 + \frac{1}{2}k(x + 0.05)^2$$

$$\Rightarrow \quad U_2 = -30(9.81)(x\sin 20°) + 10{,}000x^2 + 1{,}000x + 25 \text{ J}.$$

The energy attenuated by friction is

$$(U_{1\to2})_f = \mu_k mg(x + 8)\cos 20°$$

$$\Rightarrow \quad (U_{1\rightarrow2})_f = 0.2(30)(9.81)(x + 8)\cos 20° = 55.31(8 + x)\text{ J}.$$

Therefore,

$$U_{1\rightarrow2} = (805.254 + 25)$$
$$- \left[-30\,(9.81)\,(x\sin 20°) + 10{,}000x^2 + 1{,}000x + 25\right]$$
$$- 55.31\,(8 + x)\text{ J}.$$

Substituting all terms into Equation (4.12) and simplifying, one obtains

$$10{,}000x^2 + 954.65x - 422.72 = 0.$$

This quadratic equation gives

$$x = \frac{-954.65 \mp \sqrt{954.65^2 + 4\,(10{,}000)\,(422.72)}}{20{,}000}$$

$$\Rightarrow \quad x = \frac{-954.65 \mp \sqrt{17{,}820{,}176}}{20{,}000} = \frac{-954.65 \mp 4{,}221.395}{20{,}000} = 0.1633$$

with the negative value being disregarded. Therefore, the additional deflection is $x = 0.1633$ m.

Example 4.3
A collar of mass $m = 5$ kg is allowed to slide along the upper horizontal uniform rod, as shown in Figure 4.7a. The collar is attached to spring whose other end is anchored at a fixed point R. The spring has an undeformed length of 0.25 m and spring coefficient $k = 75$ N/m. If the collar is released from rest at P determine

(a) the speed of the collar when it reaches Q and

(b) the speed of the collar when it reaches S.

Assume that friction between the collar and rod can be disregarded.

Solution:
The given data are, $k = 75$ N/m, $m = 5$ kg, undeformed length of spring $\ell_o = 0.25$ m. Consider the length of the deformed spring at different configurations. The deformed length PR is given by

$$\ell_{PR} = \sqrt{(0.5)^2 + (0.4)^2 + (0.3)^2}\,m = 0.7071\text{ m}.$$

Similarly,

$$\ell_{QR} = \sqrt{(0.4)^2 + (0.3)^2}\,m = 0.50\text{ m},$$

$$\ell_{RS} = \sqrt{(0.5)^2 + (0.3)^2}\,m = 0.58309\text{ m}.$$

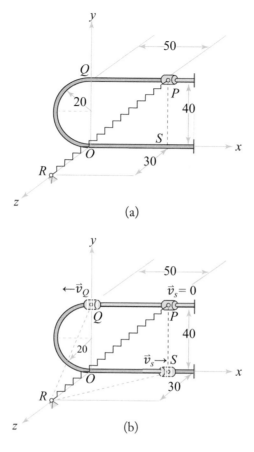

Figure 4.7: (a) Collar sliding on horizontal uniform rod and (b) different configurations of deformed spring (dimensions in cm).

Since there is no friction the principle of conservation of energy can be applied. Thus, Equation (4.8) is employed $T_1 + V_1 = T_2 + V_2$.

Speed at point Q
At starting point P, from above

$$T_P = \frac{1}{2}mv_P^2 = 0,$$

since collar starting from rest. The potential energy at P is

$$V_P = V_{Pg} + V_{Pe}$$

in which

$$V_{Pg} = mg\,(0.4m) = 19.62 \text{ N.m},$$

$$V_{Pe} = \frac{1}{2}k\left(e_{PR}^2\right) = 0.5\,(75)\,(0.45711)^2 \text{ N.m} = 7.8355 \text{ N.m},$$

in which

$$e_{PR} = \ell_{PR} - 0.250 \text{ m} = (0.7071 - 0.250)\,m = 0.45711 \text{ m}.$$

Therefore,

$$V_P = V_{Pg} + V_{Pe} = 19.62 + 7.8355 \text{ N.m} = 27.4555 \text{ N.m}.$$

At point Q,

$$T_Q = \frac{1}{2}mv_Q^2 = \frac{1}{2}\,(5)\,v_Q^2 = 2.5v_Q^2$$

while the potential energy at Q is

$$V_Q = V_{Qg} + V_{Qe}$$

in which

$$V_{Qg} = mg\,(0.4m) = V_{Pg} = 19.62 \text{ N.m},$$

$$V_{Qe} = \frac{1}{2}k\left(e_{QR}^2\right) = 0.5\,(75)\,(0.250)^2 \text{ N.m} = 2.3438 \text{ N.m},$$

$$e_{QR} = \ell_{QR} - 0.250 \text{ m} = 0.50 - 0.250 \text{ m} = 0.250 \text{ m}.$$

Therefore,

$$V_Q = V_{Qg} + V_{Qe} = 19.62 + 2.3438 \text{ N.m} = 21.9638 \text{ N.m}.$$

Substituting the foregoing results into Equation (4.8), one has

$$T_P + V_P = T_Q + V_Q$$

$$\Rightarrow \quad 0 + 27.4555 = 2.5v_Q^2 + 21.9638$$

$$\Rightarrow \quad v_Q^2 = 5.4917 \quad \Rightarrow \quad v_Q = 2.3434 \text{ m/s}.$$

Speed at point S

At starting point P, from above

$$T_P = 0, \qquad V_P = 27.4555 \text{ N.m}.$$

At point S,

$$T_S = \frac{1}{2}mv_S^2 = \frac{1}{2}\,(5)\,v_S^2 = 2.5v_S^2$$

while the potential energy at S is

$$V_S = V_{Sg} + V_{Se}$$

in which

$$V_{Sg} = mg\,(0) = 0,$$

$$V_{Se} = \frac{1}{2}k\left(e_{RS}^2\right) = 0.5(75)(0.3331)^2 = 4.1608 \text{ N.m,}$$

$$e_{RS} = \ell_{RS} - 0.250 \text{ m} = 0.58309 - 0.250 \text{ m} = 0.3331 \text{ m.}$$

Therefore,

$$V_S = V_{Sg} + V_{Se} = 4.1608 \text{ N.m.}$$

Substituting the foregoing results into Equation (4.8), one has

$$T_P + V_P = T_S + V_S$$

$$\Rightarrow \quad 0 + 27.4555 = 2.5v_S^2 + 4.1608$$

$$\Rightarrow \quad v_S^2 = 23.2947 \quad \Rightarrow \quad v_S = 4.8265 \text{ m/s.}$$

(a) The speed of the collar when it reaches Q is $v_Q = 2.3434$ m/s.

(b) The speed of the collar when it reaches S is $v_S = 4.8265$ m/s.

4.7   EXERCISES

4.1. The cable system shown in Figure 4.8 is at rest when a constant force $P = 500$ N is applied to block A. Disregarding the masses of the pulleys, effect of friction in the pulleys, and assuming that the *static* and *kinetic coefficients* of friction between block A and the surface are, respectively, $\mu_s = 0.30$ and $\mu_k = 0.25$, determine (a) velocities of blocks B and C, and (b) tension in the cable. Given that $m_A = 30$ kg, $m_B = m_C = 25$ kg.

4.2. A spring device consisting of one rigid panel connected by cables on both sides to hold on the initial 50 mm compression of the spring, as shown in Figure 4.9. If a 30 kg block having an initial velocity $v_o = 2$ m/s falling from a height of 8 m, determine the additional deflection of the spring device in bringing the block to rest. Given that the stiffness constant of the spring device is $k = 300$ kN/m.

4.3. A collar of mass $m = 5$ kg is allowed to slide along the lower horizontal uniform rod, as shown in Figure 4.10. The collar is attached to spring whose other end is anchored at a fixed point R. The spring has an undeformed length of 0.25 m and spring coefficient $k = 75$ N/m. If the collar is released from rest at P determine

(a) the speed of the collar when it reaches O, and

(b) the speed of the collar when it reaches S.

Assume that friction between the collar and rod can be disregarded.

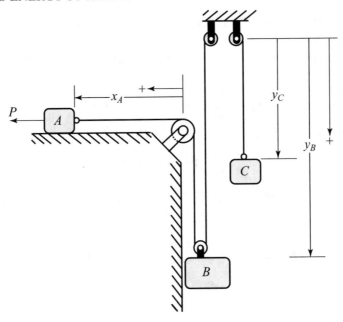

Figure 4.8: Cable system with three block masses.

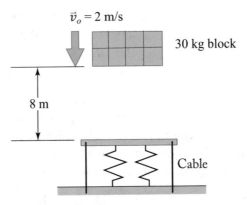

Figure 4.9: Spring device and falling block.

4.4. Two blocks B and C are suspended from an inextensible cable, as shown in Figure 4.11. If the system is released from rest, determine (a) the maximum velocity attained by block B and (b) the maximum height that block B will reach above the floor. Given that $m_B = 20$ kg, $m_C = 9$ kg.

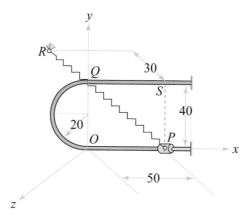

Figure 4.10: Collar sliding on rod (dimensions in mm).

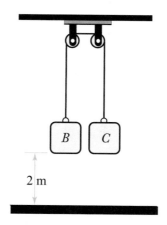

Figure 4.11: Cable system with two blocks.

4.5. The system shown in Figure 4.12 is at rest when a force $P = 50$ N is applied to block A. If the surfaces are all smooth such that friction can be disregarded, determine the velocity of block A after it has moved 1 m. Given that $m_A = 20$ kg, $m_C = 20$ kg.

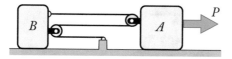

Figure 4.12: Two sliding blocks on smooth surface.

4.6. Repeat the problem in Figure 4.12 above but assume now that the coefficients of static and kinetic friction are, $\mu_s = 0.30$ and $\mu_k = 0.25$, respectively.

4.7.   A spring device consisting of one rigid panel connected by cables on both sides to hold on the initial 60 mm compression of the spring, as shown in Figure 4.13. If an 80 kg block having an initial velocity $v_o = 2$ m/s moving to the right to further compress 20 mm of the spring device, find (a) the coefficient of kinetic friction between the block and the horizontal surface and (b) the velocity of the block as it returns to the original position shown in the figure after compressing the spring device. Given that the stiffness constant of the spring device is $k = 300$ kN/m.

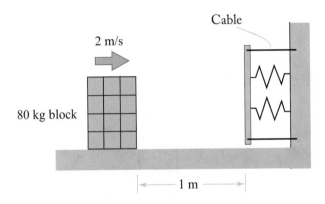

Figure 4.13: Spring device and moving block.

4.8.   A loop *ABCDE*, as shown in Figure 4.14, is on a smooth horizontal table (not shown). A small ball of mass $m = 0.1$ kg is released from rest at *A* when the spring is compressed 30 mm and it travels around the loop. Determine the smallest value of the spring constant for which the small ball will travel around the loop and will remain in contact with the loop at all the time.

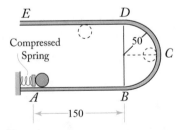

Figure 4.14: Ball travels in loop *ABCDE* (dimensions in mm).

CHAPTER 5

Impulse, Momentum, and Impact of Particles

5.1 INTRODUCTION

In the previous chapters two basic methods for the solution of problems of particles have been introduced. These methods are as follows.

- Direct application of Newton's second law of motion,

$$\vec{F} = m\vec{a}.$$

  In this method the principles of kinematics are applied such that one can find position, velocity from $\vec{a}$ at any time.

- Method of work and energy which relates force, mass, velocity, and displacement.

  In this chapter a third basic method is presented. This is the *method of impulse and momentum*. Note that in this method and the method of *work and energy* the determination of acceleration $\vec{a}$ is unnecessary.

  In the next section, Section 5.2, the principle of impulse and momentum is introduced. Impulsive motion and impact are dealt with in Section 5.3. The main focus is on direct central impact, oblique central impact, and constrained oblique impact. Consideration of energy and momentum is included in Section 5.4.

5.2 PRINCIPLE OF IMPULSE AND MOMENTUM

Consider Newton's second law of motion

$$\vec{F} = m\vec{a} = \frac{d(m\vec{v})}{dt},$$

where $m\vec{v}$ is the linear momentum and m is constant. One can multiply both sides of this equation by dt and integrating between times t_1 and t_2 such that

$$\int_{t_1}^{t_2} \vec{F}\, dt = \int_{\vec{v}_1}^{\vec{v}_2} d(m\vec{v}) = m\vec{v}_2 - m\vec{v}_1.$$

Rearranging, one has

$$m\vec{v}_1 + \int_{t_1}^{t_2} \vec{F}\, dt = m\vec{v}_2.$$

Note that the integral on the lhs is known as the *impulse in the interval between t_1 and t_2*. The above equation can be written as

$$m\vec{v}_1 + I_{1\to 2} = m\vec{v}_2, \tag{5.1}$$

where the impulse in the time interval, $I_{1\to 2} = \int_{t_1}^{t_2} \vec{F}\, dt$ has been applied.

Equation (5.1) states that the *final momentum $m\vec{v}_2$ of a particle acting on by an impulse of the force $\vec{F}$ is equal to the vector sum of its initial momentum and the impulse of the force $\vec{F}$ during the interval of time in question.*

Equation (5.1) can diagrammatically be expressed as in Figure 5.1.

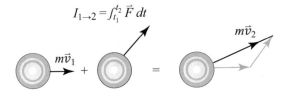

Figure 5.1: Illustration of Equation (5.1).

When a particle is acted on by several forces the impulse of every force has to be included so that by applying Equation (5.1) one obtained

$$m\vec{v}_1 + \sum I_{1\to 2} = m\vec{v}_2. \tag{5.2}$$

When the problem involves two or n particles of mass m_i, Equation (5.1) can be applied separately to every particle. One can also express

$$\sum_{i=1}^{n} m_i \left(\vec{v}_i\right)_1 + \sum_{i=1}^{n} \int_{t_1}^{t_2} \vec{F}_i\, dt = \sum_{i=1}^{n} m_i \left(\vec{v}_i\right)_2 \qquad \text{or}$$

$$\sum_{i=1}^{n} m_i \left(\vec{v}_i\right)_1 + \sum_{i=1}^{n} \left(I_i\right)_{1\to 2} = \sum_{i=1}^{n} m_i \left(\vec{v}_i\right)_2. \tag{5.3}$$

In applying this equation one should note that only the impulses of external forces have to be considered since the forces of action and reaction exerted by the particles on one another form pairs of equal and opposite forces which cancel out. If no external force is acted on the particles or if the vector sum of the external forces is zero the second term on the lhs of Equation (5.3) vanishes. Thus, Equation (5.3) reduces to

$$\sum_{i=1}^{n} m_i \left(\vec{v}_i \right)_1 = \sum_{i=1}^{n} m_i \left(\vec{v}_i \right)_2.$$ (5.4)

Equation (5.4) states that *the total momentum of the particles is conserved.*

More on impulse and momentum of a system of particles are presented in Chapter 6.

5.3   IMPULSIVE MOTION AND IMPACT

- An *impulsive force* is defined as one acting upon a particle during a very short duration which is significant enough to produce a definite change in momentum. The resulting motion is known as the *impulsive motion.*

When impulsive forces exert on a particle application of Equation (5.2) gives

$$m \vec{v}_1 + \sum \vec{F} \Delta t = m \vec{v}_2.$$ (5.5)

Of course, in the cases in which several particles with impulsive forces are involved, Equations (5.3) and (5.4) can be applied.

- *Impact* is defined as a collision between two bodies (conceptually, they are considered as particles) which occurs in a small interval of time.

- *Line of impact* is defined as the common normal to the surfaces in contact during the impact.

- When the mass centers of the bodies are located on the *line of impact* and the velocities are along the line of impact such an impact is called *central impact*, as shown in Figure 5.2a. When the velocities are *not* along the line of impact such an impact is referred to as *oblique central impact*, as shown in Figure 5.2b.

5.3.1   DIRECT CENTRAL IMPACT

Consider two particles (for example, two spherical balls) A and B of mass m_A and m_B, respectively, which are moving in a straight line as in Figure 5.3.

Note that in Figure 5.3 the velocity $\vec{v}_A$ has to be larger than $\vec{v}_B$, otherwise no impact is possible.

By the conservation of momentum and with reference to the symbols in Figure 5.3, one can write

$$m_A \vec{v}_A + m_B \vec{v}_B = m_A \vec{v}'_A + m_B \vec{v}'_B,$$ (5.6)

where $\vec{v}'_A$ and $\vec{v}'_B$ are, respectively, the velocities of m_A and m_B after the impact.

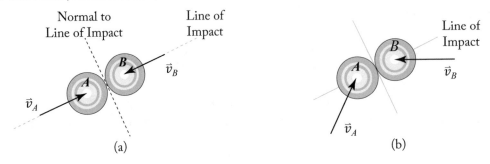

Figure 5.2: (a) Direct central impact, and (b) oblique central impact.

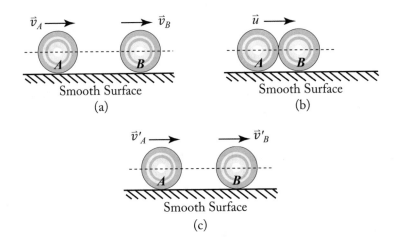

Figure 5.3: (a) Before impact (b) at maximum deformation, and (c) after impact.

In this particular case all particles are moving in the same line therefore one can remove the vector notation. Thus, the scalar equation becomes

$$m_A v_A + m_B v_B = m_A v'_A + m_B v'_B. \tag{5.7}$$

This equation has two unknowns which are the velocities after impact v'_A and v'_B.

In order to solve for the two unknowns one needs to bring in another equation. To this end one can consider the actions during the impact. These actions take place during the *period of deformation* and *period of restitution*, as diagrammatically illustrated in Figure 5.4.

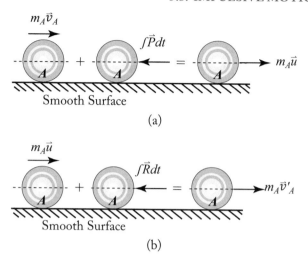

Figure 5.4: Actions during impact: (a) period of deformation and (b) period of restitution.

Considering particle A and applying the *principle of impulse and momentum* to the period of deformation and period of restitution, one has

$$m_A v_A - \int P \, dt = m_A u, \qquad (5.8)$$

$$m_A u - \int R \, dt = m_A v'_A \qquad (5.9)$$

in which the negative sign on the lhs indicates the direction is in opposite to that of v_A.

From Equation (5.8), one has

$$\int P \, dt = m_A v_A - m_A u. \qquad (5.10)$$

From Equation (5.9), one obtains

$$\int R \, dt = m_A u - m_A v'_A. \qquad (5.11)$$

However, the *coefficient of restitution*, e is defined as

$$e = \frac{\int R \, dt}{\int P \, dt}.$$

Substituting from Equations (5.10) and (5.11), one has

$$e = \frac{\int R \, dt}{\int P \, dt} = \frac{m_A u - m_A v'_A}{m_A v_A - m_A u} = \frac{u - v'_A}{v_A - u}. \qquad (5.12)$$

Similarly, by considering particle B, one can show

$$e = \frac{v_B' - u}{u - v_B}. \tag{5.13}$$

From Equations (5.12) and (5.13) the unknown velocity u is to be eliminated.

From Equation (5.12), one obtains

$$e(v_A - u) = u - v_A'.$$

Rearranging terms to obtain u,

$$u = \frac{ev_A + v_A'}{1 + e}. \tag{5.14}$$

From Equation (5.13), one has

$$u = \frac{ev_B + v_B'}{1 + e}. \tag{5.15}$$

Equating (5.14) to (5.15), it gives

$$ev_A + v_A' = ev_B + v_B'.$$

Rearranging terms, one has

$$e = \frac{v_B' - v_A'}{v_A - v_B}. \tag{5.16}$$

Equation (5.16) expresses the coefficient of restitution as a ratio of the relative velocity after impact to that before impact of the two particles. Applying Equation (5.7) and (5.16) the two unknown velocities after impact, v_A' and v_B', can be solved.

Remarks:

(a) For perfectly plastic impact, $e = 0$.

(b) For perfectly elastic impact, $e = 1$. Thus, in this case, the total energy as well as the total momentum of the two particles is conserved. That is, the kinetic energy of the two particles is conserved

$$\frac{1}{2}m_A v_A^2 + \frac{1}{2}m_B v_B^2 = \frac{1}{2}m_A \left(v_A'\right)^2 + \frac{1}{2}m_B \left(v_B'\right)^2. \tag{5.17}$$

5.3.2   OBLIQUE CENTRAL IMPACT

Now, consider two colliding particles whose velocities are not directed along the *line of impact*, as shown in Figure 5.2b. This is the case known as *oblique central impact*. Suppose the velocities $\vec{v}_A'$ and $\vec{v}_B'$ after impact are in the directions shown in Figure 5.5. These unknown vector quantities constitute four unknowns, two for unknown magnitudes and two for unknown angles or directions. Their solution requires four independent equations.

Assume coordinate n is along the line of impact and coordinate t, perpendicular to n, is the common tangent. If the surfaces of the particles are perfectly *smooth* and *frictionless* as well as *no external force* applied to the particles, the only impulses acted on the particles during the impact are the internal forces directed along the line of impact. That is, the internal forces are along the n-coordinate in Figure 5.5. This means that the impact of the two particles has the following three situations.

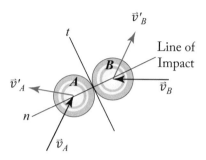

Figure 5.5: Oblique central impact of two particles.

- First, the component along the t-axis of the momentum of each particle is conserved. That is, the t component of the velocity of each particle remains unchanged. This means that

$$\left(\vec{v}_A\right)_t = \left(\vec{v}'_A\right)_t ,$$ (5.18a)
$$\left(\vec{v}_B\right)_t = \left(\vec{v}'_B\right)_t$$ (5.18b)

  in which the subscript t denotes the t component of the velocity.

- Second, the component along the n-axis of the two momenta of the two particles is conserved. This enables one to write

$$m_A \left(\vec{v}_A\right)_n + m_B \left(\vec{v}_B\right)_n = m_A \left(\vec{v}'_A\right)_n + m_B \left(\vec{v}'_B\right)_n .$$ (5.19)

- Third, the component along the n-axis of the relative velocity of the two particles after impact can be obtained by multiplying the coefficient of restitution and the relative velocity before impact. This is similar to that for the direct central impact. This gives

$$\left(\vec{v}'_B\right)_n - \left(\vec{v}'_A\right)_n = e\left[\left(\vec{v}_A\right)_n - \left(\vec{v}_B\right)_n\right].$$ (5.20)

The four independent equations, Equations (5.18a), (5.18b), (5.19), and (5.20), can be applied to solve for the four unknowns (two unknown magnitudes and two unknown phases or angles) of the velocities $\vec{v}'_A$ and $\vec{v}'_B$ of the two particles after impact.

5.3.3   CONSTRAINED OBLIQUE CENTRAL IMPACT

In the foregoing sections, the two particles are assumed to be moving freely before and after impact. In this section it is assumed that one or both of the colliding particles is or are constrained. Specifically, consider the case illustrated in Figure 5.6 in which block A is constrained to move horizontally on the smooth surface and ball B is free to move in the plane of the figure. For simplicity, it is assumed that no friction between block A and the horizontal surface, and no friction between block A and ball B. It should be noted that the impulses acted on the system consist of those of the internal forces $\vec{F}$ and $-\vec{F}$ directed along the line of impact (that is, the n-axis), and the impulse of the external force $\vec{F}_e$ acted on by the horizontal surface on block A and directed along the vertical.

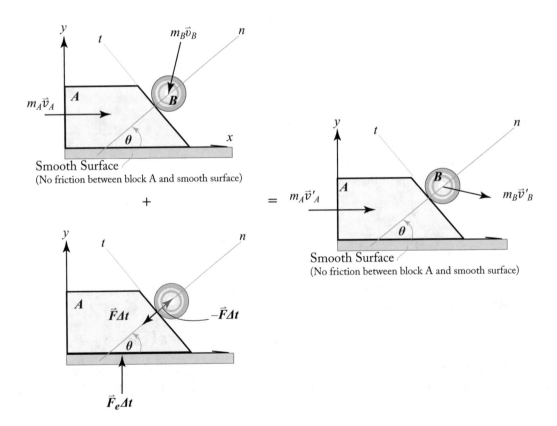

Figure 5.6: Constrained oblique central impact of a block and a ball.

The unknown velocities of block A and ball B immediately after impact are the magnitude of the velocity $\vec{v}'_A$ (its direction is known to be along the horizontal surface), the magnitude and direction of $\vec{v}'_B$. This means that one requires three independent equations to solve for the three unknowns. To this end, one considers the following three situations.

- First, since there is no friction between block A and ball B, the component along the t-axis of the momentum of ball B is conserved. Thus, the t component of the velocity of ball B remains unaltered. This enables one to write

$$\left(\vec{v}_B\right)_t = \left(\vec{v}'_B\right)_t. \tag{5.21}$$

- Second, the component along the x-axis of the total momentum of block A and ball B is conserved. This means

$$m_A v_A + m_B \left(v_B\right)_x = m_A v'_A + m_B \left(v'_B\right)_x. \tag{5.22}$$

- Third, the component along the n-axis of the relative velocity of block A and ball B after impact can be obtained by multiplying the coefficient of restitution and the n component of the relative velocity before impact. That is

$$\left(\vec{v}'_B\right)_n - \left(\vec{v}'_A\right)_n = e\left[\left(\vec{v}_A\right)_n - \left(\vec{v}_B\right)_n\right]. \tag{5.23}$$

It may be appropriate to note that the validity of Equation (5.23) cannot be established via a simple extension of the derivation for the direct central impact case. Without presenting the proof it suffices to state that Equation (5.23) is valid to cases in which the motion of one particle or both particles are constrained.

Equations (5.21), (5.22), and (5.23) can be applied to solve for the three unknowns (two unknown magnitudes and one unknown phase or direction) of the velocities $\vec{v}'_A$ and $\vec{v}'_B$ after impact.

5.4   CONSIDERATION OF ENERGY AND MOMENTUM

Up to this stage three basic methods for the solution of problems of particles have been introduced. These methods are as follows.

- Direct application of Newton's second law of motion,

$$\vec{F} = m\vec{a}.$$

In this method the principles of kinematics are applied such that one can find position, velocity from $\vec{a}$ at any time.

- Method of work and energy which relates force, mass, velocity, and displacement.

- Method of impulse and momentum which has been presented in this chapter.

Of course, the choice of one or more of these methods is solely dependent of the type of problem one is confronted with. For example, there are situations in which the solution of the problem at hand may require the combined use of the foregoing methods. Specifically, for example, in the case of two pendulums P and Q are hanging from the ceiling. When mass m_P of pendulum P is released from its horizontal position P_1 to hit mass m_Q of pendulum Q at rest, as illustrated in Figure 5.7a, the solution of this problem requires the following three phases.

(a) In the first phase, *pendulum P travels from position P_1 to P_2*, as shown in Figure 5.7a. The principle of conservation of energy can be applied to find the velocity $(\vec{v}_P)_2$ of pendulum P at position P_2.

(b) In the second phase, *pendulum P impacts pendulum Q*, as shown in Figure 5.7b. Since there is no externally applied impulse the total momentum of the two pendulums is conserved. Further, the velocities $(\vec{v}_P)_3$ and $(\vec{v}_Q)_3$ of the two pendulums after impact can be determined by applying the relation between their relative velocities.

(c) In the third and final phase, *pendulum Q travels from position Q_3 to Q_4*, as shown in Figure 5.7c. The principle of conservation of energy can be applied to find the maximum vertical distance y_4 that can be achieved by pendulum Q. The corresponding angle θ can thus be found by trigonometry.

Example 5.1

A much simplified model of explosion of a grenade conceptually consists of two hemispheres that are connected by an inextensible string holding the spring under compression, as shown in Figure 5.8a in which $m_A = 2.0$ kg, $m_B = 1.0$ kg. Note that the spring is not attached to the hemispheres that have unequal masses. It is known that the potential energy of the compressed spring is 100 J or N.m and the whole system has an initial velocity $\vec{v}_o$. When the string is cut, which simulates the explosion event, the hemispheres fly apart and the angle between the axis of the system and the horizontal is $\theta = 61°$. If the magnitude of the initial velocity is $v_o = 100$ m/s, determine the resulting velocity of each hemisphere.

Solution:

Let the frame Oxy moving with the mass center (z-axis being perpendicular to the Oxy plane) while the x-axis and y-axis along the horizontal and vertical directions with the origin O at the mass center, as indicated in Figure 5.8b. Note that since the frame is moving with the whole system therefore, the velocity of each hemisphere is zero before the string is cut. This means that the kinetic energy, immediately after the string is cut, is zero.

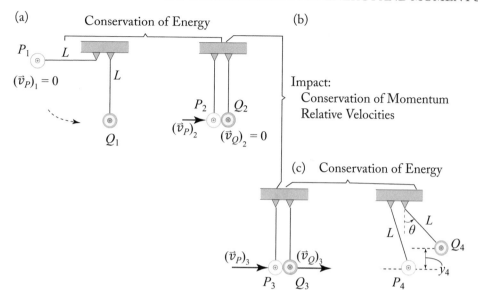

Figure 5.7: Three phases in the solution of two impacting pendulums: (a) conservation of energy in first phase, (b) impact of two pendulums, and (c) conservation of energy in third phase.

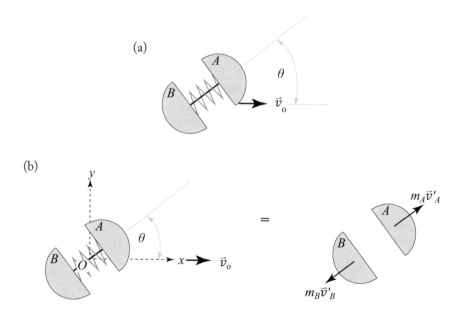

Figure 5.8: (a) Model of explosion of a grenade and (b) FBD with frame Oxy.

Given data

$m_A = 2.0$ kg, $m_B = 1.0$ kg, $V_1 = 100$ N.m, and $V_2 = 0$, because there is no strain energy when the string is cut.

Conservation of momentum

$$0 + 0 = m_A \vec{v}'_A - m_B \vec{v}'_B.$$

Note that the negative sign on the rhs indicates that immediately after the string is cut the two hemispheres travel in opposite directions. Therefore,

$$\vec{v}'_A = \left(\frac{m_B}{m_A}\right)\vec{v}'_B.$$

Conservation of energy

$$T_1 + V_1 = T_2 + V_2$$

$$\Rightarrow \quad 0 + 100 = \left[\frac{1}{2}m_A \left(v'_A\right)^2 + \frac{1}{2}m_B \left(v'_B\right)^2\right] + 0$$

$$\Rightarrow \quad 100 = \frac{1}{2}m_A \left[\left(\frac{m_B}{m_A}\right)\vec{v}'_B\right]^2 + \frac{1}{2}m_B \left(v'_B\right)^2$$

$$\Rightarrow \quad 100 = \left[\frac{m_B \left(m_A + m_B\right)}{2m_A}\right]\left(v'_B\right)^2$$

$$\Rightarrow \quad v'_B = \sqrt{\frac{200 m_A}{m_B \left(m_A + m_B\right)}}.$$

Substituting for the given data, one obtains

$$v'_B = \sqrt{\frac{200\,(2)}{1\,(2+1)}}\,\frac{m}{s} = 20\sqrt{\frac{1}{3}}\,\frac{m}{s}, \quad v'_A = 40\sqrt{\frac{1}{3}}\,\frac{m}{s}.$$

$$\vec{v}'_A = 40\sqrt{\frac{1}{3}}\,\frac{m}{s} \; \angle\; 61°, \quad \vec{v}'_B = 20\sqrt{\frac{1}{3}}\,\frac{m}{s} \; \measuredangle\; 61°.$$

Velocities of A and B

The absolute velocities of A and B are

$$\left(\vec{v}'_A\right)_F = [100 \text{ m/s} \rightarrow] + \left[40\sqrt{\frac{1}{3}}\,\frac{m}{s} \; \angle\; 61°\right] = 113.02\,\frac{m}{s} \; \angle\; 10.29°,$$

$$\left(\vec{v}'_B\right)_F = [100 \text{ m/s} \rightarrow] + \left[20\sqrt{\frac{1}{3}}\,\frac{m}{s} \; \measuredangle\; 61°\right] = 94.94\,\frac{m}{s} \; \measuredangle\; 6.07°.$$

The detailed steps of obtaining these absolute velocities are provided in the following.

Determination of $\left(\vec{v}'_A\right)_F$ and $\left(\vec{v}'_B\right)_F$

Recall the Oxy moves with the mass center. The unit vectors along the x-axis and y-axis are, respectively, $\vec{\imath}$ and $\vec{\jmath}$ as shown in Figure 5.8b. Thus,

$$\left(\vec{v}'_A\right)_F = \vec{v}_o + \vec{v}'_A = 100\vec{\imath} + 40\sqrt{\frac{1}{3}}\left(\cos 61°\,\vec{\imath} + \sin 61°\,\vec{\jmath}\right)\frac{m}{s}$$

$$\Rightarrow \left(\vec{v}'_A\right)_F = \left(100 + 40\sqrt{\frac{1}{3}}\cos 61°\right)\vec{\imath} + \left(40\sqrt{\frac{1}{3}}\sin 61°\right)\vec{\jmath}\,\frac{m}{s}$$

$$\Rightarrow \left(\vec{v}'_A\right)_F = 111.20\vec{\imath} + 20.20\vec{\jmath}\ \text{m/s} \quad \text{or} \quad \left(\vec{v}'_A\right)_F = 113.02\ \text{m/s} \quad \angle\ 10.29°.$$

Similarly,

$$\left(\vec{v}'_B\right)_F = \vec{v}_o + \vec{v}'_B = 100\vec{\imath} - 20\sqrt{\frac{1}{3}}\left(\cos 61°\vec{\imath} + \sin 61°\vec{\jmath}\right)\frac{m}{s}$$

$$\Rightarrow \left(\vec{v}'_B\right)_F = \left(100 - 20\sqrt{\frac{1}{3}}\cos 61°\right)\vec{\imath} - \left(20\sqrt{\frac{1}{3}}\sin 61°\right)\vec{\jmath}\,\frac{m}{s}$$

$$\Rightarrow \left(\vec{v}'_B\right)_F = 94.40\vec{\imath} - 10.10\vec{\jmath}\ \text{m/s} \quad \text{or} \quad \left(\vec{v}'_B\right)_F = 94.94\ \text{m/s} \quad \angle\ 6.07°.$$

Example 5.2

A spherical ball B is attached at the free end of an inextensible cable which is fixed to the ceiling, as shown in Figure 5.9a. Another spherical ball A having the same density and half of the radius of ball B is released from rest when it is just touching the cable and acquires a velocity $\vec{v}_o$ before striking ball B. Assuming the coefficient of restitution $e = 1$ and friction between the surfaces of the balls being disregarded, find the velocity of each ball immediately after impact.

Solution:

Let the mass of ball B be $m_B = m$ and its radius r. Then mass of ball A is $m_A = m/8$ since volume of ball B is $V_B = \frac{4}{3}\pi r^3$ and radius of ball A is $r/2$.

For ball A (Figure 5.9b)

Momentum is conserved along tangent to balls. By applying Equation (5.2), one has

$$\frac{m}{8}\vec{v}_A + I_{1\to 2} = \frac{m}{8}\vec{v}'_A.$$

Components along $t - t$,

$$+\searrow, \qquad \frac{m}{8}v_o \sin\theta + 0 = \frac{m}{8}\left(v'_A\right)_t \quad \Rightarrow \quad \left(v'_A\right)_t = v_o \sin\theta$$

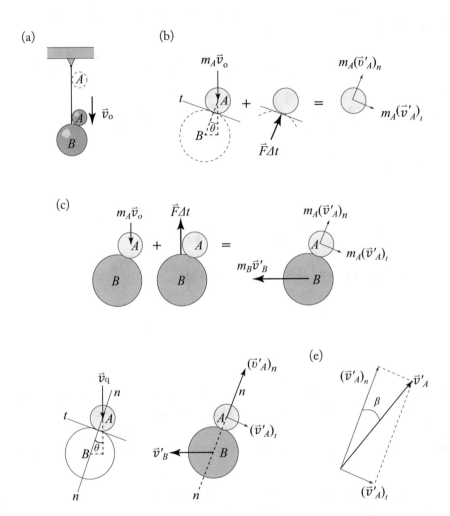

Figure 5.9: (a) Two spherical balls undergoing impact, (b) FBD for principle of impulse and momentum, (c) FBD for conservation of total momentum, (d) relative velocities along n-n, and (e) direction of $\vec{v}'_A$.

where $\sin\theta = \frac{r/2}{r+r/2} = \frac{1}{3} \Rightarrow \theta = 19.47°$. Thus, $\left(v'_A\right)_t = \frac{v_o}{3}$.

For balls A and B (Figure 5.9c)
Total momentum is conserved. That is,

$$\frac{m}{8}\vec{v}_A + \vec{I}_{1\to2} = \frac{m}{8}\vec{v}'_A + m\vec{v}'_B.$$

Components along horizontal x-axis,

$$+x \rightarrow, \qquad 0 + 0 = \frac{m}{8}\left(v'_A\right)_t \cos\theta + \frac{m}{8}\left(v'_A\right)_n \sin\theta - mv'_B$$

$$\Rightarrow \quad 0 = \frac{m}{8}\left(\frac{v_o}{3}\right)\left(\frac{2\sqrt{2}}{3}\right) + \frac{m}{8}\left(v'_A\right)_n\left(\frac{1}{3}\right) - mv'_B$$

$$\Rightarrow \quad \left(v'_A\right)_n\left(\frac{1}{24}\right) - v'_B + \left(\frac{\sqrt{2}}{36}\right)v_o = 0. \qquad (5.24a)$$

For relative velocities along n-n (Figure 5.9d)
With reference to Figure 5.9d and the given condition that $e = 1$, one can write

$$\left(v'_A\right)_n - \left(v'_B\right)_n = (v_B)_n - (v_A)_n.$$

But $\left(v'_B\right)_n = -v'_B \sin\theta$, $(v_B)_n = 0$, $(v_A)_n = -v_o \cos\theta$, and $\sin\theta = \frac{1}{3}$, $\cos\theta = \frac{2\sqrt{2}}{3}$. Therefore,

$$\left(v'_A\right)_n + \frac{v'_B}{3} = \frac{2\sqrt{2}}{3}v_o \quad \Rightarrow \quad \left(v'_A\right)_n = \frac{2\sqrt{2}}{3}v_o - \frac{v'_B}{3}. \qquad (5.24b)$$

Substituting Equation (5.24b) into (5.24a),

$$\left(\frac{2\sqrt{2}}{3}v_o - \frac{v'_B}{3}\right)\left(\frac{1}{24}\right) - v'_B + \left(\frac{\sqrt{2}}{36}\right)v_o = 0$$

$\Rightarrow \quad v'_B = \left(\frac{4\sqrt{2}}{73}\right)v_o = 0.0775v_o$. That is, $\vec{v}'_B = 0.0775v_o \quad \leftarrow$. Substituting this result into Equation (5.24b),

$$\left(v'_A\right)_n = \frac{2\sqrt{2}}{3}v_o - \frac{1}{3}\left(\frac{4\sqrt{2}}{73}\right)v_o = 0.9170v_o.$$

Therefore,

$$v'_A = \sqrt{0.9170^2 + 0.3333^2}\,v_o = 0.9757v_o.$$

With reference to Figure 5.9e,

$$\tan\beta = \frac{\left(v'_A\right)_t}{\left(v'_A\right)_n} = \frac{0.3333v_o}{0.9170v_o} = 0.3635 \quad \Rightarrow \quad \beta = 19.97°.$$

Therefore, the angle of $\vec{v}_A'$ with the horizontal x-axis is

$$90° - \theta - \beta = 90° - 19.47° - 19.97° = 50.56°.$$

Thus,

$$\vec{v}_A' = 0.9757 v_o \quad \measuredangle\ 50.56° \quad \text{and}$$
$$\vec{v}_B' = 0.0775 v_o \quad \leftarrow.$$

Example 5.3

A pendulum with the spherical bob of mass $m_A = 1.0$ kg has a speed $\vec{v}_o = 2$ m/s, as shown in Figure 5.10a. The inextensible length of the rope is $\ell = 1.0$ m. The angle between the vertical direction and the rope is $\beta = 45°$. Wedge B is confined to move along the horizontal direction without friction between the ball-bearings and surface. It is connected to a spring whose stiffness constant is $k = 1,000$ N/m and the angle $\theta = 30°$. If the mass of B is $m_B = 1.5$ kg and the coefficient of restitution between A and B is $e = 0.8$, determine the velocities of A and B immediately after impact.

Solution:

The principle of conservation of energy can be applied to the event just before the impact.

Given data

$$m_A = 1.0 \text{ kg}, \quad m_B = 1.5 \text{ kg}, \quad v_o = 2 \text{ m/s}, \quad k = 1,000 \text{ N/m},$$
$$e = 0.8, \quad \beta = 45°, \quad \theta = 30°, \quad \ell = 1.0 \text{ m}.$$

For bob A

At the initial state, the kinetic energy is

$$T_o = \frac{1}{2} m_A v_o^2 = \frac{1}{2}(1)(2)^2 \text{ J} = 2 \text{ J}.$$

The potential energy is

$$V_o = m_A g h_o, \quad h_o = \ell(1 - \cos 45°) = 1.0\,(1 - 0.7071) \text{ m} = 0.2929 \text{ m}.$$

Therefore,

$$V_o = (1.0)\,(9.81)\,(0.2929) \text{ J} = 2.8733 \text{ J}.$$

Just before impact,

$$\beta = 0, \quad h_1 = 0, \quad V_1 = 0, \quad T_1 = \frac{1}{2} m_A v_A^2 = 0.5 v_A^2.$$

Applying the principle of conservation of energy, $T_o + V_o = T_1 + V_1$. Therefore,

$$2 + 2.8733 = 0 + 0.5 v_A^2$$

which gives

$$v_A = 3.1220 \text{ m/s} \quad \text{or} \quad \vec{v}_A = 3.1220 \text{ m/s} \quad \rightarrow.$$

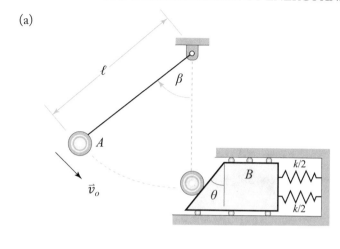

(a)

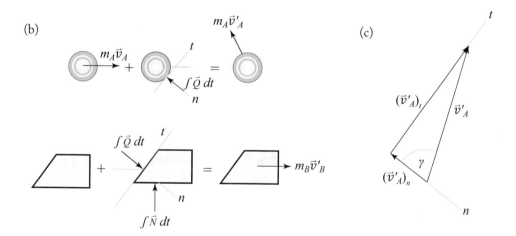

(b)

(c)

Figure 5.10: (a) Impact between the pendulum and the wedge, (b) FBD for use of conservation of momentum, and (c) direction of $\vec{v}'_A$.

Impact analysis (Figure 5.10b)

Since there is no friction between the bob and the wedge, therefore the principle of conservation of momentum can be employed in the present analysis. It should be noted that since the rope becomes slack during impact no impulse is presented in the rope.

For spherical bob A

Momentum in the t-direction,

$$m_A v_A \sin \theta + 0 = m_A \left(v'_A \right)_t$$

$$\Rightarrow \quad (v_A')_t = v_A \sin 30° = (3.1220)(.50) \text{ m/s} = 1.5610 \text{ m/s}$$

$$\Rightarrow \quad \left(\vec{v}_A'\right)_t = 1.5610 \text{ m/s} \quad \measuredangle \; 60°.$$

For spherical bob A and wedge B
Momentum in horizontal or x-axis,

$$m_A v_A + 0 = m_A \left(v_A'\right)_n \cos\theta + m_A \left(v_A'\right)_t \sin\theta + m_B v_B'$$

$$\Rightarrow \quad v_A = \left(v_A'\right)_n \cos 30° + \left(v_A'\right)_t \sin 30° + (1.5)v_B'$$

$$\Rightarrow \quad \left(v_A'\right)_n \cos 30° + (1.5)\, v_B' = 3.1220 - (1.5610)(0.5)$$

$$\Rightarrow \quad \left(v_A'\right)_n \cos 30° + (1.5)\, v_B' = 2.3415. \tag{5.25a}$$

For relative velocities along n-n
Applying the definition of the coefficient of restitution, one has

$$\left(v_B'\right)_n - \left(v_A'\right)_n = e\left[(v_A)_n - (v_B)_n\right]$$

$$\Rightarrow \quad v_B' \cos\theta - \left(v_A'\right)_n = e\left[v_A \cos\theta - 0\right]$$

$$\Rightarrow \quad v_B' \cos 30° - \left(v_A'\right)_n = (0.8)(3.1220)\cos 30°. \tag{5.25b}$$

Solving Equations (5.25a) and (5.25b) simultaneously, one obtains

$$\left(v_A'\right)_n = -0.5408 \text{ m/s}, \quad v_B' = 1.8732 \text{ m/s}.$$

Direction of velocity $\vec{v}_A'$
Magnitude of $\vec{v}_A'$ is given by

$$v_A' = \sqrt{\left(v_A'\right)_t^2 + \left(v_A'\right)_n^2} = \sqrt{(1.5610)^2 + (-0.5408)^2} \text{ m/s}$$

$$\Rightarrow \quad v_A' = 1.6520 \text{ m/s}.$$

With reference to Figure 5.10c,

$$\tan\gamma = \frac{\left(v_A'\right)_t}{-\left(v_A'\right)_n} = \frac{1.5610}{0.5408} \quad \Rightarrow \quad \gamma = 70.88°.$$

That is, $\theta + \gamma = 100.88°$. Therefore,

$$\vec{v}_A' = 1.6520 \text{ m/s} \quad \diagdown \; 100.88°,$$

and

$$\vec{v}_B' = 1.8732 \text{ m/s} \quad \rightarrow.$$

5.5 EXERCISES

5.1. A 1.0 kg ball A is moving with a velocity $\vec{v}_A$ of magnitude 10 m/s, as shown in Figure 5.11. It is hit by a 2 kg ball B having a velocity $\vec{v}_B$ of magnitude 8 m/s. If friction can be disregarded and the coefficient of restitution is 0.6, find the velocity of each ball after impact.

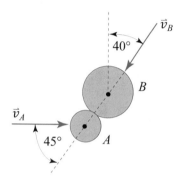

Figure 5.11: Impacting balls.

5.2. A 10 kg cylinder A rests on a 3 kg platform B supported by a cable which passes over the pulleys C and D to end at the block E, as shown in Figure 5.12. The mass of block E is $m_E = 5$ kg. If the system is released from rest, determine (a) the velocity of block E after 0.6 s and (b) the force exerted by the cylinder on the platform.

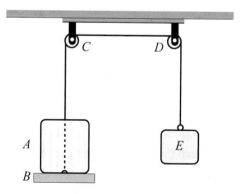

Figure 5.12: Cylinder rests on platform with counter block.

5.3. Three homogeneous identical spherical balls are suspended from the ceiling by inextensible cables of equal length L which are spaced at a distance slightly greater than the diameter of the balls, as shown in Figure 5.13 (this is commercially known as Newton's cradle). If ball A is pulled back at the position A_o and released such that it hits ball B which then hits ball C. Given that the coefficient of restitution is e and $\vec{v}_o$ the velocity of ball A just before it hits B, determine (a) the velocities of A and B immediately after the first collision and (b) the velocities of B and C immediately after the second collision.

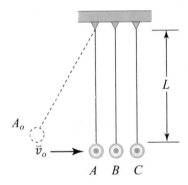

Figure 5.13: Newton's cradle.

5.4. In a billiard ball game ball A is hit so that its velocity $\vec{v}_A$ of magnitude 5 m/s is moving in the direction shown in Figure 5.14 when it strikes ball B that is at rest. The coefficient of restitution is 0.8 while the diameter of each of the two balls is 60 mm. If after impact ball B is moving in the x-direction, determine (a) the angle θ and (b) the velocity of ball B after impact.

5.5. A pendulum with the spherical bob of mass $m_A = 2.0$ kg has a speed $\vec{v}_o = 5$ m/s, as shown in Figure 5.15. The inextensible length of the rope is $\ell = 2.0$ m. The angle between the vertical direction and the rope is $\beta = 45°$. Block B is confined to move along the horizontal direction without friction between the ball-bearings and surface. It is connected to two identical springs whose combined stiffness constant is $k = 2{,}000$ N/m. If the mass of block B is $m_B = 5$ kg and the coefficient of restitution between A and B is $e = 0.8$, determine the velocities of A and B immediately after impact.

5.6. The collar cable block system shown in Figure 5.16 consists of a collar A of mass $m_A = 30$ kg and a counter block B of mass $m_A = 25$ kg is at rest when a constant force F of 200 N is applied to collar A. By disregarding friction and mass of pulleys, determine

(a) the speed of collar A just before it hits the floor C, and

(b) the speed of collar A just before it hits the floor C if block B is replaced by a downward force of 25 N.

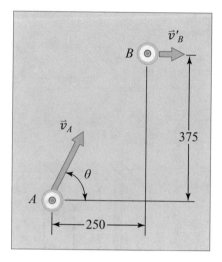

Figure 5.14: Two billiard balls in motion (dimensions in mm).

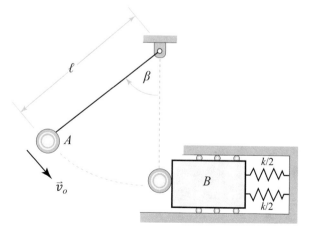

Figure 5.15: Impact between a pendulum and block of mass.

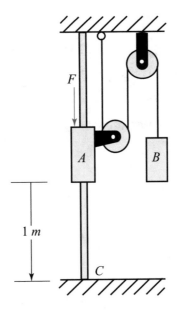

Figure 5.16: Collar cable block system.

C H A P T E R 6

Systems of Particles

6.1   INTRODUCTION

In the previous chapters, three basic methods for the solution of problems of particles have been introduced. These methods are: (a) the method of direct application of Newton's second law of motion, (b) the method of work and energy, and (c) the method of impulse and momentum. In the present chapter the aforementioned three basic methods are generalized for applications to problems involving systems of particles.

The presentation of this chapter is as what follows. Newton's laws of motion for systems of particles are presented in Section 6.2. Linear and angular momenta of a system of particles are dealt with in Section 6.3. Motion of mass center of a system of particles is introduced in Section 6.4. Section 6.5 is concerned with the angular momentum of a system of particles about its mass center. Conservation of linear and angular momentum for a system of particles is studied in Section 6.6. Work energy principle is introduced in Section 6.7 while conservation of energy for a system of particles is included in Section 6.8. The principle of impulse and momentum for a system of particles is considered in Section 6.9.

6.2   NEWTON'S LAWS OF MOTION FOR SYSTEMS OF PARTICLES

In order to set up the equations of motion for a system of n particles one applies Newton's second law of motion to every particle. Consider one of these particles, P_i as shown in Figure 6.1 where the number of particles in the system is n, the subscript $i = 1, 2, \ldots, n$, m_i is the mass of particle P_i, $\vec{a}_i$ is its acceleration w.r.t. the Newtonian or fixed frame of reference (*NFR* or *FFR*) *OXYZ*, $\vec{F}_i$ is the resultant of all the *external forces* exerting on particle P_i, and $\vec{f}_{ij}$ is the *internal* force of P_i acted on by another particle P_j. With reference to Figure 6.1 and applying Newton's second law of motion, one has

$$\vec{F}_i + \sum_{j=1}^{n} \vec{f}_{ij} = m_i \vec{a}_i, \tag{6.1}$$

where the term on the rhs is known as the effective force of particle P_i. It should be noted that when $i = j$, $\vec{f}_{ij} = 0$ since it has no meaning.

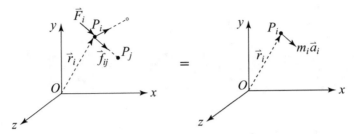

Figure 6.1: Particle P_i in a system of particles.

Taking the moments about the origin of *FFR*, one obtains

$$\vec{r}_i \times \vec{F}_i + \vec{r}_i \times \sum_{j=1}^{n} \vec{f}_{ij} = \vec{r}_i \times m_i \vec{a}_i \tag{6.2}$$

in which $\vec{r}_i$ is the position vector of particle P_i.

Similarly, summing forces of all the particles in the system according to Equation (6.1), one arrives at

$$\sum_{i=1}^{n} \vec{F}_i + \sum_{i=1}^{n}\sum_{j=1}^{n} \vec{f}_{ij} = \sum_{i=1}^{n} m_i \vec{a}_i \tag{6.3}$$

and summing moments of all the particles in the system with reference to Equation (6.2), one obtains

$$\sum_{i=1}^{n} \vec{r}_i \times \vec{F}_i + \sum_{i=1}^{n} \vec{r}_i \times \sum_{j=1}^{n} \vec{f}_{ij} = \sum_{i=1}^{n} \vec{r}_i \times m_i \vec{a}_i \quad \text{or}$$

$$\sum_{i=1}^{n} \vec{r}_i \times \vec{F}_i + \sum_{i=1}^{n}\sum_{j=1}^{n} \vec{r}_i \times \vec{f}_{ij} = \sum_{i=1}^{n} \vec{r}_i \times m_i \vec{a}_i. \tag{6.4}$$

But according to Newton's third law of motion,

$$\vec{f}_{ij} + \vec{f}_{ji} = 0, \tag{6.5}$$

and the sum of their moments about origin O is

$$\vec{r}_i \times \vec{f}_{ij} + \vec{r}_j \times \vec{f}_{ji} = \vec{r}_i \times \left(\vec{f}_{ij} + \vec{f}_{ji} \right) + (\vec{r}_j - \vec{r}_i) \times \vec{f}_{ji},$$

where the first term on the rhs vanishes due to Newton's third law of motion and the second term on the rhs is equal to zero because $\vec{r}_j - \vec{r}_i$ and $\vec{f}_{ji}$ are collinear, as shown in Figure 6.2. Therefore,

$$\vec{r}_i \times \vec{f}_{ij} + \vec{r}_j \times \vec{f}_{ji} = 0. \tag{6.6}$$

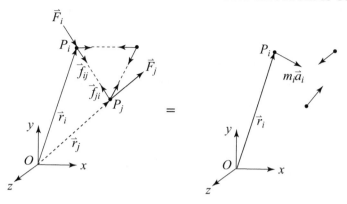

Figure 6.2: Collinear vectors $\vec{r}_j - \vec{r}_i$ and $\vec{f}_{ji}$.

Adding all the internal forces of the system,

$$\sum_{i=1}^{n}\sum_{j=1}^{n}\vec{f}_{ij} = 0 \tag{6.7}$$

and summing moments about O of these internal forces

$$\sum_{i=1}^{n}\sum_{j=1}^{n}\vec{r}_i \times \vec{f}_{ij} = 0. \tag{6.8}$$

Equations (6.7) and (6.8) are simply the expressions of the fact that the resultant and moment resultant of the internal forces of the system are zero.

By making use of Equation (6.7), Equation (6.3) reduces to

$$\sum_{i=1}^{n}\vec{F}_i = \sum_{i=1}^{n}m_i\vec{a}_i. \tag{6.9}$$

and applying Equation (6.8), Equation (6.4) becomes

$$\sum_{i=1}^{n}\vec{r}_i \times \vec{F}_i = \sum_{i=1}^{n}\vec{r}_i \times m_i\vec{a}_i. \tag{6.10}$$

Equations (6.9) and (6.10) state the fact that *the system of external forces* $\vec{F}_i$ *and the system of the effective forces* $m_i\vec{a}_i$ *possess the same resultant and the same moment resultant.*

6.3   LINEAR AND ANGULAR MOMENTUM OF A SYSTEM OF PARTICLES

Another approach for solution to the system of particles is to relate the *effective forces* of the particles to their linear momenta. Writing the linear momentum of a system of particles as

$$\vec{L} = \sum_{i=1}^{n} m_i \vec{v}_i. \tag{6.11}$$

In addition, one requires the angular momentum $\vec{H}_O$ about the origin O of the system of particles. By making use of the definition of angular momentum of a particle in Section 3.6, one obtains

$$\vec{H}_O = \sum_{i=1}^{n} \vec{r}_i \times m_i \vec{v}_i. \tag{6.12}$$

Taking the time derivative of Equation (6.11),

$$\dot{\vec{L}} = \frac{d\vec{L}}{dt} = \sum_{i=1}^{n} m_i \frac{d\vec{v}_i}{dt} = \sum_{i=1}^{n} m_i \vec{a}_i. \tag{6.13}$$

Comparing the rhs of Equations (6.9) with that of the last equation, one concludes that the rate of change of the linear momentum is equal to the effective forces of the system,

$$\dot{\vec{L}} = \sum_{i=1}^{n} \vec{F}_i = \sum_{i=1}^{n} m_i \vec{a}_i. \tag{6.14}$$

Taking the time derivative of Equation (6.12),

$$\dot{\vec{H}}_O = \frac{d\vec{H}_O}{dt} = \frac{d\left(\sum_{i=1}^{n} \vec{r}_i \times m_i \vec{v}_i\right)}{dt} = \sum_{i=1}^{n} \left(\frac{d\vec{r}_i}{dt} \times m_i \vec{v}_i + \vec{r}_i \times m_i \frac{d\vec{v}_i}{dt}\right)$$

and since the first term inside the brackets on the rhs vanishes due to the fact that vector cross-product of the vector itself is zero, the above equation becomes

$$\dot{\vec{H}}_O = \sum_{i=1}^{n} \vec{r}_i \times m_i \frac{d\vec{v}_i}{dt} = \sum_{i=1}^{n} \vec{r}_i \times m_i \vec{a}_i = \sum_{i=1}^{n} \left(\vec{M}_O\right)_i, \tag{6.15}$$

where $\left(\vec{M}_O\right)_i$ is the moment of the particle P_i about origin O.

Thus, Equations (6.14) and (6.15) state that *the rates of change of the linear momentum and of the angular momentum about fixed origin O of the external forces are respectively equal to the force resultant and moment resultant about origin O of the system of particles.*

6.4   MOTION OF MASS CENTER OF A SYSTEM OF PARTICLES

Equation (6.14) can be expressed in a different form when one considers the mass center of a system of particles. To begin, one defines

$$m\overline{\vec{r}} = \sum_{i=1}^{n} m_i \vec{r}_i, \tag{6.16}$$

where $\overline{\vec{r}}$ is the position vector between the fixed origin O and the mass center G of the system of particles, as shown in Figure 6.3 while the total mass $m = \sum_{i=1}^{n} m_i$.

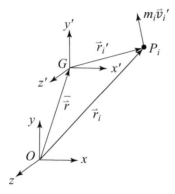

Figure 6.3: Mass center and its position vector.

With reference to Figure 6.3,

$$\vec{r}_i = \overline{\vec{r}} + \vec{r}_i' \tag{6.17}$$

in which $\vec{r}_i'$ is the position vector between the mass center G and particle P_i.

Substituting Equation (6.17) into (6.16) and recall that $m = \sum_{i=1}^{n} m_i$, one has

$$m\overline{\vec{r}} = \sum_{i=1}^{n} m_i \left(\overline{\vec{r}} + \vec{r}_i' \right) = \sum_{i=1}^{n} m_i \overline{\vec{r}} + \sum_{i=1}^{n} m_i \vec{r}_i' = m\overline{\vec{r}} + \sum_{i=1}^{n} m_i \vec{r}_i'$$

which leads to

$$\sum_{i=1}^{n} m_i \vec{r}_i' = 0. \tag{6.18}$$

This equation simply states the fact that G is the mass center of the system of particles. It explains why

$$\sum_{i=1}^{n} m_i \vec{r}_i' \times \overline{\vec{a}} = 0, \tag{6.19}$$

and

$$\sum_{i=1}^{n} m_i \vec{r}_i' \times \vec{a}_i \neq 0. \tag{6.20}$$

Equations (6.19) and (6.20) can better be understood by considering a system of three particles. Thus, the lhs of Equation (6.19) gives

$$\sum_{i=1}^{3} m_i \vec{r}_i' \times \vec{\bar{a}} = \left(m_1 \vec{r}_1' + m_2 \vec{r}_2' + m_3 \vec{r}_3' \right) \times \vec{\bar{a}}, \tag{6.21}$$

and the lhs of Equation (6.20) becomes

$$\sum_{i=1}^{3} m_i \vec{r}_i' \times \vec{a}_i = m_1 \vec{r}_1' \times \vec{a}_1 + m_2 \vec{r}_2' \times \vec{a}_2 + m_3 \vec{r}_3' \times \vec{a}_3. \tag{6.22}$$

By applying Equation (6.18), the rhs of Equation (6.21) is equal to zero whereas the rhs of Equation (6.22), in general, is not equal to zero.

Now, time differentiating both sides of Equation (6.16), one has

$$m \frac{d\vec{\bar{r}}}{dt} = m \dot{\vec{\bar{r}}} = \sum_{i=1}^{n} m_i \frac{d\vec{r}_i}{dt} = \sum_{i=1}^{n} m_i \dot{\vec{r}}_i \quad \text{or} \quad m \vec{\bar{v}} = \sum_{i=1}^{n} m_i \vec{v}_i, \tag{6.23}$$

where $\vec{\bar{v}}$ denotes the velocity of the mass center G of the system of particles. The lhs of Equation (6.23) is, by definition, the linear momentum $\vec{L}$ of the system and therefore, taking the time derivative of Equation (6.23), one has

$$\dot{\vec{L}} = m \vec{\bar{a}} = \sum_{i=1}^{n} m_i \vec{a}_i, \tag{6.24}$$

where $\vec{\bar{a}}$ is the acceleration of the mass center G of the system of particles.

By making use of the definitions of $\vec{v}_i = \vec{\bar{v}} + \vec{v}_i'$ and $\vec{a}_i = \vec{\bar{a}} + \vec{a}_i'$, and substituting these into Equations (6.23) and (6.24), respectively, one has

$$m \vec{\bar{v}} = \sum_{i=1}^{n} m_i \vec{v}_i = \sum_{i=1}^{n} m_i \left(\vec{\bar{v}} + \vec{v}_i' \right) = \sum_{i=1}^{n} m_i \vec{\bar{v}} + \sum_{i=1}^{n} m_i \vec{v}_i',$$

and

$$m \vec{\bar{a}} = \sum_{i=1}^{n} m_i \vec{a}_i = \sum_{i=1}^{n} m_i \left(\vec{\bar{a}} + \vec{a}_i' \right) = \sum_{i=1}^{n} m_i \vec{\bar{a}} + \sum_{i=1}^{n} m_i \vec{a}_i'.$$

But $\sum_{i=1}^{n} m_i \, \overline{\vec{v}} = m \overline{\vec{v}}$ and $\sum_{i=1}^{n} m_i \, \overline{\vec{a}} = m \overline{\vec{a}}$, therefore one obtains

$$\sum_{i=1}^{n} m_i \, \vec{v}_i' = 0, \tag{6.25a}$$

$$\sum_{i=1}^{n} m_i \, \vec{a}_i' = 0. \tag{6.25b}$$

6.5   ANGULAR MOMENTUM OF A SYSTEM OF PARTICLES ABOUT ITS MASS CENTER

In Section 6.3, the linear and angular momentum of a system of particles has been expressed w.r.t. the *FFR*. However, in some situations it is more convenient to consider the angular momentum w.r.t. the mass center G of the system of particles, as shown in Figure 6.4.

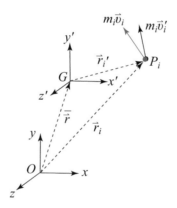

Figure 6.4: Fixed frame of reference and mass center.

With reference to Figure 6.4 in which a frame of reference with G being the origin is shown, the angular momentum $\vec{H}_G'$ of the system of particles about the mass center of a system of particles is defined by

$$\vec{H}_G' = \sum_{i=1}^{n} \vec{r}_i' \times m_i \, \vec{v}_i'. \tag{6.26}$$

Taking the time derivative of $\vec{H}_G'$ gives

$$\dot{\vec{H}}_G' = \frac{d \vec{H}_G'}{dt} = \frac{d \left(\sum_{i=1}^{n} \vec{r}_i' \times m_i \, \vec{v}_i' \right)}{dt} = \sum_{i=1}^{n} \left(\frac{d \vec{r}_i'}{dt} \times m_i \, \vec{v}_i' + \vec{r}_i' \times m_i \, \frac{d \vec{v}_i'}{dt} \right)$$

in which the first term inside the brackets on the rhs vanishes due to the fact that vector cross-product of the vector itself is zero. Thus, this equation reduces to

$$\dot{\vec{H}}'_G = \sum_{i=1}^{n} \vec{r}'_i \times m_i \frac{d\vec{v}'_i}{dt} = \sum_{i=1}^{n} \vec{r}'_i \times m_i \vec{a}'_i. \tag{6.27}$$

Now, taking the second time derivative of Equation (6.17), one obtains

$$\vec{a}_i = \bar{\vec{a}} + \vec{a}'_i$$

so that upon substituting this equation into Equation (6.27), one has

$$\dot{\vec{H}}'_G = \sum_{i=1}^{n} \vec{r}'_i \times m_i \left(\vec{a}_i - \bar{\vec{a}} \right) = \sum_{i=1}^{n} \vec{r}'_i \times m_i \vec{a}_i - \left(\sum_{i=1}^{n} \vec{r}'_i \times m_i \right) \bar{\vec{a}}.$$

By making use of Equation (6.19) the second term on the rhs is zero, and therefore this equation reduces to

$$\dot{\vec{H}}'_G = \sum_{i=1}^{n} \vec{r}'_i \times m_i \vec{a}_i = \sum_{i=1}^{n} \vec{r}'_i \times \vec{F}_i = \sum_{i=1}^{n} \left(\vec{M_G} \right)_i, \tag{6.28}$$

where $\vec{F}_i = m_i \vec{a}_i$ has been applied and $\left(\vec{M_G} \right)_i$ is the moment of the particle P_i about origin G. Thus, Equation (6.28) simply states that *the sum of all the moments about G of all particles in the system is equal to the rate of change of the angular momentum about G of the system of particles.*

It is interesting to note that by taking the first time derivative of Equation (6.17) and substituting the resulting equation into Equation (6.26), one can show that

$$\vec{H}'_G = \sum_{i=1}^{n} \vec{r}'_i \times m_i \vec{v}'_i = \sum_{i=1}^{n} \vec{r}'_i \times m_i \vec{v}_i = \vec{H}_G. \tag{6.29}$$

Equation (6.29) simply states that *the sum of all the angular momenta about G based on the velocities of the particles with reference to G is identical to that of all angular momenta about G based on the velocities of the particles with reference to the origin O of the FFR.*

Furthermore, with reference to Figure 6.4, the sum of angular momenta of all particles about O is

$$\vec{H}_O = \sum_{i=1}^{n} \vec{r}_i \times m_i \vec{v}_i = \sum_{i=1}^{n} \left(\bar{\vec{r}} + \vec{r}'_i \right) \times m_i \vec{v}_i,$$

by Equation (6.17).

Expanding,

$$\vec{H}_O = \sum_{i=1}^{n} \left(\bar{\vec{r}} \times m_i \vec{v}_i \right) + \sum_{i=1}^{n} \vec{r}'_i \times m_i \vec{v}_i$$

$$\Rightarrow \quad \vec{H}_O = \bar{\vec{r}} \times \sum_{i=1}^{n} m_i \vec{v}_i + \sum_{i=1}^{n} \vec{r}_i' \times m_i \vec{v}_i.$$

By Equation (6.23), the first summation term on the rhs becomes

$$\bar{\vec{r}} \times \sum_{i=1}^{n} m_i \vec{v}_i = \bar{\vec{r}} \times \left(m \bar{\vec{v}} \right),$$

and the second summation term, by Equation (6.29), is $\vec{H}_G$. Therefore, one has the following important result,

$$\vec{H}_O = \bar{\vec{r}} \times \left(m \bar{\vec{v}} \right) + \vec{H}_G. \tag{6.30}$$

6.6 CONSERVATION OF MOMENTUM FOR A SYSTEM OF PARTICLES

Now, returning to Equations (6.14) and (6.15), one observes that if there is no external force acting on the particles of a system, which means that $\sum_{i=1}^{n} \vec{F}_i = 0$ and $\sum_{i=1}^{n} \left(\vec{M}_o \right)_i = 0$, then one has

$$\dot{\vec{L}} = 0, \tag{6.31a}$$

$$\dot{\vec{H}}_o = 0. \tag{6.31b}$$

Upon integrating w.r.t time t, it becomes

$$\vec{L} = constant, \tag{6.32a}$$

$$\vec{H}_o = constant. \tag{6.32b}$$

These two equations simply state that *the momentum of the system of particles and the angular momentum of the same system of particles about the origin O are conserved.*

6.7 WORK ENERGY PRINCIPLE FOR A SYSTEM OF PARTICLES

The work energy principle for a particle stated in Equation (4.6) can be applied to every particle in the system of particles. Then the form of the work energy principle for a system of particles is the same as that in Equation (4.6). Of course, now

$$T_1 = \frac{1}{2} \sum_{i=1}^{n} m_i \left(v_i^2 \right)_1, \quad T_2 = \frac{1}{2} \sum_{i=1}^{n} m_i \left(v_i^2 \right)_2,$$

and

$$U_{1\to2} = \sum_{i=1}^{n} \int_{s_1}^{s_2} (F_t)_i \, ds$$

in which $(F_t)_i$ represents the tangential component of the resultant force acting on particle P_i by several forces. That is, explicitly, the work energy principle for a system of particles becomes

$$\frac{1}{2} \sum_{i=1}^{n} m_i \left(v_i^2\right)_1 + \sum_{i=1}^{n} \int_{s_1}^{s_2} (F_t)_i \, ds = \frac{1}{2} \sum_{i=1}^{n} m_i \left(v_i^2\right)_2$$

or in more concise form

$$T_1 + U_{1\to2} = T_2. \tag{6.33}$$

6.8   CONSERVATION OF ENERGY FOR A SYSTEM OF PARTICLES

Since the kinetic energy of a system of particles is given by

$$T = \frac{1}{2} \sum_{i=1}^{n} m_i v_i^2.$$

This can be expressed in terms of velocity at the mass center G and velocities reference to this mass center. To this end, one first considers the velocity $\vec{v}_i$ of particle P_i,

$$\vec{v}_i = \bar{\vec{v}} + \vec{v}_i',$$

as illustrated in Figure 6.4 in which $\bar{\vec{v}}$ is the velocity of the mass center G.

Note that $\vec{v}_i \cdot \vec{v}_i = v_i^2$. Therefore, the kinetic energy of a system of particles becomes

$$T = \frac{1}{2} \sum_{i=1}^{n} m_i v_i^2 = \frac{1}{2} \sum_{i=1}^{n} m_i \left(\bar{\vec{v}} + \vec{v}_i'\right) \cdot \left(\bar{\vec{v}} + \vec{v}_i'\right)$$

$$\Rightarrow \quad T = \frac{1}{2} \left(\sum_{i=1}^{n} m_i\right) \bar{v}^2 + \bar{\vec{v}} \cdot \sum_{i=1}^{n} m_i \vec{v}_i' + \frac{1}{2} \sum_{i=1}^{n} m_i \left(v_i'\right)^2.$$

According to Equation (6.25a), the second summation term on the rhs is zero. Therefore, the kinetic energy of the system of particles reduces to

$$T = \frac{1}{2} m \bar{v}^2 + \frac{1}{2} \sum_{i=1}^{n} m_i \left(v_i'\right)^2. \tag{6.34}$$

The principle of conservation of energy expressed in Equation (4.8) for a particle can be applied to a system of particles. Thus, if the forces acting on the particles are *conservative*, Equation (6.33) can be written as

$$T_1 + V_1 = T_2 + V_2, \tag{6.35}$$

where V_1 and V_2 are the potential energies contributed by all the internal and external forces acting on the particles of the system at times t_1 and t_2. Equation (6.35) is known as the *principle of conservation of energy for a system of particles.*

6.9 PRINCIPLE OF IMPULSE AND MOMENTUM FOR A SYSTEM OF PARTICLES

In Section 6.6, the conservation of momentum for a system of particles has been introduced. This is based on the assumption that there is no external force exerting on the particles. In general, when the external forces are not zero and upon integrating Equation (6.14) w.r.t. time t, one has

$$\int_{t_1}^{t_2} \dot{\vec{L}} \, dt = L_2 - L_1 = \sum_{i=1}^{n} \int_{t_1}^{t_2} \vec{F}_i \, dt \quad \text{or} \quad L_1 + \sum_{i=1}^{n} \int_{t_1}^{t_2} \vec{F}_i \, dt = L_2. \tag{6.36}$$

This equation is known as the *principle of impulse and momentum for a system of particles*. It simply states that for a system of particles *the sum of the linear momentum at instant t_1 and all the impulses in the duration between t_1 and t_2 is equal to the linear momentum at instant t_2.*

Similarly, for the case in which external angular momenta are not zero and upon integrating Equation (6.15) w.r.t. time t, one obtains

$$\int_{t_1}^{t_2} \dot{\vec{H}}_O \, dt = (H_O)_2 - (H_O)_1 = \sum_{i=1}^{n} \int_{t_1}^{t_2} \left(\vec{M}_O \right)_i dt$$

or

$$(H_O)_1 + \sum_{i=1}^{n} \int_{t_1}^{t_2} \left(\vec{M}_O \right)_i dt = (H_O)_2. \tag{6.37}$$

Equation (6.37) is referred to as the *principle of angular impulse and angular momentum for a system of particles about the origin O of the FFR.* It simply states that for a system of particles *the sum of the angular momentum about the origin O at instant t_1 and all the angular impulses about O in the duration between t_1 and t_2 is equal to the angular momentum about O at instant t_2.*

Example 6.1
A system of three particles A, B, and C, shown in Figure 6.5, has the following masses and position vectors:

$$m_A = 2.0 \text{ kg}, \qquad m_B = 3.0 \text{ kg}, \qquad m_C = 4.0 \text{ kg},$$

$$\vec{r}_A = -3\vec{i} + 3\vec{j} + 6\vec{k} \text{ m}, \qquad \vec{r}_B = -3\vec{i} + 4\vec{j} + 2\vec{k} \text{ m},$$

and

$$\vec{r}_C = 2\vec{i} - 3\vec{j} - 4\vec{k} \text{ m}.$$

The velocities of these particles are given as, $\vec{v}_A = -5\vec{i} + 4\vec{j} + 3\vec{k}$ m/s, $\vec{v}_B = 3\vec{i} + 2\vec{j} + 1\vec{k}$ m/s, and $\vec{v}_C = 6\vec{i} - 5\vec{j} - 4\vec{k}$ m/s. Determine

(a) the position vector $\vec{r}$ of the mass center G of the system,

(b) the linear momentum $m\vec{v}$ of the system,

(c) the angular momentum $\vec{H}_G$ of the system, and

(d) show that Equation (6.30) is true.

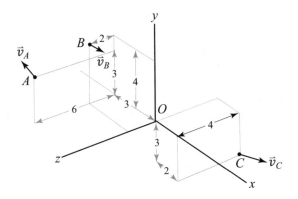

Figure 6.5: System of three particles (dimension in m).

Solution:
Given masses,

$$m_A = 2.0 \text{ kg}, \qquad m_B = 3.0 \text{ kg}, \qquad m_C = 4.0 \text{ kg}.$$

Given position vectors,

$$\vec{r}_A = -3\vec{i} + 3\vec{j} + 6\vec{k} \text{ m}, \quad \vec{r}_B = -3\vec{i} + 4\vec{j} + 2\vec{k} \text{ m}, \quad \vec{r}_C = 2\vec{i} - 3\vec{j} - 4\vec{k} \text{ m}.$$

Given velocity vectors,

$$\vec{v}_A = -5\vec{i} + 4\vec{j} + 3\vec{k} \text{ m/s}, \qquad \vec{v}_B = 3\vec{i} + 2\vec{j} + 1\vec{k} \text{ m/s},$$

$$\vec{v}_C = 6\vec{i} - 5\vec{j} - 4\vec{k} \text{ m/s}.$$

(a) *Position vector of mass center*

$$(m_A + m_B + m_C)\bar{\vec{r}} = m_A\vec{r}_A + m_B\vec{r}_B + m_C\vec{r}_C$$

$$(2 + 3 + 4)\bar{\vec{r}} = 2\left(-3\vec{i} + 3\vec{j} + 6\vec{k}\right) + 3\left(-3\vec{i} + 4\vec{j} + 2\vec{k}\right)$$

$$+4\left(2\vec{i} - 3\vec{j} - 4\vec{k}\right)$$

$$9\bar{\vec{r}} = \left(-6\vec{i} + 6\vec{j} + 12\vec{k}\right) + \left(-9\vec{i} + 12\vec{j} + 6\vec{k}\right)$$

$$+\left(8\vec{i} - 12\vec{j} - 16\vec{k}\right)$$

$$9\bar{\vec{r}} = -7\vec{i} + 6\vec{j} + 2\vec{k} \quad \Rightarrow \quad \bar{\vec{r}} = \frac{1}{9}\left(-7\vec{i} + 6\vec{j} + 2\vec{k}\right) \text{ m.}$$

(b) *Linear momentum of system* $m\bar{\vec{v}}$

Linear momentum of every particle,

$$m_A\vec{v}_A = 2\left(-5\vec{i} + 4\vec{j} + 3\vec{k}\right) \text{ kg.m/s} = -10\vec{i} + 8\vec{j} + 6\vec{k} \text{ kg.m/s}$$

$$m_B\vec{v}_B = 3\left(3\vec{i} + 2\vec{j} + 1\vec{k}\right) \text{ kg.m/s} = 9\vec{i} + 6\vec{j} + 3\vec{k} \text{ kg.m/s}$$

$$m_C\vec{v}_C = 4\left(6\vec{i} - 5\vec{j} - 4\vec{k}\right) \text{ kg.m/s} = 24\vec{i} - 20\vec{j} - 16\vec{k} \text{ kg.m/s.}$$

Linear momentum of system is

$$m\bar{\vec{v}} = m_A\vec{v}_A + m_B\vec{v}_B + m_C\vec{v}_C = 23\vec{i} - 6\vec{j} - 7\vec{k} \text{ kg.m/s.}$$

(c) *Angular momentum about* **G**

Position vectors relative to the mass center,

$$\vec{r}'_A = \vec{r}_A - \bar{\vec{r}} = -3\vec{i} + 3\vec{j} + 6\vec{k} - \frac{1}{9}\left(-7\vec{i} + 6\vec{j} + 2\vec{k}\right) \text{ m}$$

$$= \frac{1}{9}\left(-20\vec{i} + 21\vec{j} + 52\vec{k}\right) \text{ m}$$

$$\vec{r}'_B = \vec{r}_B - \bar{\vec{r}} = -3\vec{i} + 4\vec{j} + 2\vec{k} - \frac{1}{9}\left(-7\vec{i} + 6\vec{j} + 2\vec{k}\right) \text{ m}$$

$$= \frac{1}{9}\left(-20\vec{i} + 30\vec{j} + 16\vec{k}\right) \text{ m}$$

$$\vec{r}'_C = \vec{r}_C - \bar{\vec{r}} = 2\vec{i} - 3\vec{j} - 4\vec{k} - \frac{1}{9}\left(-7\vec{i} + 6\vec{j} + 2\vec{k}\right) \text{ m}$$

$$= \frac{1}{9}\left(25\vec{i} - 33\vec{j} - 38\vec{k}\right) \text{ m.}$$

Angular momentum about G,

$$\vec{H}_G = \vec{r}'_A \times m_A \vec{v}_A + \vec{r}'_B \times m_B \vec{v}_B + \vec{r}'_C \times m_C \vec{v}_C.$$

But

$$\vec{r}'_A \times m_A \vec{v}_A = \begin{vmatrix} \vec{i} & \vec{j} & \vec{k} \\ -20 & 21 & 52 \\ \dfrac{}{9} & \dfrac{}{9} & \dfrac{}{9} \\ -10 & 8 & 6 \end{vmatrix} = -\frac{290}{9}\vec{i} - \frac{400}{9}\vec{j} + \frac{50}{9}\vec{k} \text{ kg.m}^2/\text{s.}$$

$$\vec{r}'_B \times m_B \vec{v}_B = \begin{vmatrix} \vec{i} & \vec{j} & \vec{k} \\ -20 & 30 & 16 \\ \dfrac{}{9} & \dfrac{}{9} & \dfrac{}{9} \\ 9 & 6 & 3 \end{vmatrix} = -\frac{6}{9}\vec{i} + \frac{204}{9}\vec{j} - \frac{390}{9}\vec{k} \text{ kg.m}^2/\text{s.}$$

$$\vec{r}'_C \times m_C \vec{v}_C = \begin{vmatrix} \vec{i} & \vec{j} & \vec{k} \\ 25 & -33 & -38 \\ \dfrac{}{9} & \dfrac{}{9} & \dfrac{}{9} \\ 24 & -20 & -16 \end{vmatrix} = -\frac{232}{9}\vec{i} - \frac{512}{9}\vec{j} + \frac{292}{9}\vec{k} \text{ kg.m}^2/\text{s.}$$

Summing, one has

$$\vec{H}_G = -\frac{528}{9}\vec{i} - \frac{708}{9}\vec{j} - \frac{48}{9}\vec{k} \text{ kg.m}^2/\text{s.} \tag{6.38a}$$

(d) **Angular momentum about O**

By Equation (6.30), $\vec{H}_O = \vec{r} \times \left(m\vec{v} \right) + \vec{H}_G$ where the first term on the rhs

$$\vec{r} \times m\vec{v} = \begin{vmatrix} \vec{i} & \vec{j} & \vec{k} \\ -7 & 6 & 2 \\ \dfrac{}{9} & \dfrac{}{9} & \dfrac{}{9} \\ 23 & -6 & -7 \end{vmatrix} = -\frac{30}{9}\vec{i} - \frac{3}{9}\vec{j} - \frac{96}{9}\vec{k} \text{ kg.m}^2/\text{s.} \tag{6.38b}$$

Summing Equations (6.38a) and (6.38b),

$$\vec{H}_O = -62\vec{i} - 79\vec{j} - 16\vec{k} \text{ kg.} \frac{\text{m}^2}{\text{s}}. \tag{6.38c}$$

Now, applying the definition of angular momentum of a system of particles about O,

$$\vec{H}_O = \vec{r}_A \times m_A \vec{v}_A + \vec{r}_B \times m_B \vec{v}_B + \vec{r}_C \times m_C \vec{v}_C. \tag{6.38d}$$

But

$$\vec{r}_A \times m_A \vec{v}_A = \begin{vmatrix} \vec{i} & \vec{j} & \vec{k} \\ -3 & 3 & 6 \\ -10 & 8 & 6 \end{vmatrix} = -30\vec{i} - 42\vec{j} + 6\vec{k} \text{ kg.m}^2/\text{s,}$$

$$\vec{r}_B \times m_B \vec{v}_B = \begin{vmatrix} \vec{i} & \vec{j} & \vec{k} \\ -3 & 4 & 2 \\ 9 & 6 & 3 \end{vmatrix} = 0 + 27\vec{j} - 54\vec{k} \ \text{kg.m}^2/\text{s},$$

$$\vec{r}_C \times m_C \vec{v}_C = \begin{vmatrix} \vec{i} & \vec{j} & \vec{k} \\ 2 & -3 & -4 \\ 24 & -20 & -16 \end{vmatrix} = -32\vec{i} - 64\vec{j} + 32\vec{k} \ \text{kg.m}^2/\text{s}.$$

Therefore, Equation (6.38d) becomes

$$\vec{H}_O = -30\vec{i} - 42\vec{j} + 6\vec{k} + 27\vec{j} - 54\vec{k}$$
$$- 32\vec{i} - 64\vec{j} + 32\vec{k} \ \text{kg.m}^2/\text{s}.$$
$$= -62\vec{i} - 79\vec{j} - 16\vec{k} \ \text{kg.m}^2/\text{s}.$$

This is identical to that in Equation (6.38c). Thus,

$$\vec{H}_O = \bar{\vec{r}} \times m\bar{\vec{v}} + \vec{H}_G.$$

That is, Equation (6.30) is true.

Example 6.2

Two micro-satellites A and B of mass $m_A = 100$ kg and $m_B = 80$ kg, respectively, are in orbit and connected by an inextensible, taut high carbon cable of length $\ell = 10$ m, as diagrammatically shown in Figure 6.6a. The satellites spin counter-clockwise (c.c.) about their mass center G at a rate of $\omega = 20$ rad/s. At $t = 0$, the coordinates of G are $\bar{x}_o = 0$, $\bar{y}_o = 10$ m, and its velocity is $\vec{v}_o = 50\vec{i} + 10\vec{j}$ m/s. Subsequently (to the extent that these given data have not changed), the cable breaks apart, satellite A appears to move along a trajectory parallel to the y-axis while satellite B along a trajectory that intersects the horizontal x-axis at a distance $h_B = 80$ m from the origin O. Disregarding gravitational forces and other forces, such as those due to solar winds and impacts of space debris, determine

(a) the velocities of the satellites immediately after the cable breaks, and

(b) the distance h_A from y-axis to satellite A.

Solution:

Let the coordinates of satellite A be x_A and y_A while the coordinates of satellite B, x_B, and y_B.

Given data,

$$m_A = 100 \text{ kg}, \quad m_B = 80 \text{ kg}, \quad AG = r'_A, \quad BG = r'_B,$$

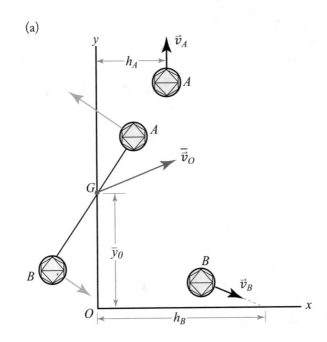

(a)

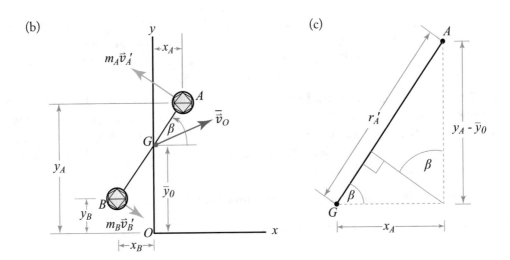

Figure 6.6: (a) Two satellites in orbit and positions after cable is broken, (b) satellites with New-tonian frame and angular momentum, and (c) sketch illustrating mass center and r'_A.

$$h_B = 80 \text{ m}, \quad AB = \ell = 10 \text{ m},$$

$$\bar{x}_o = 0, \quad \bar{y}_o = 10 \text{ m}, \quad \omega = 20 \text{ rad/s}, \quad \vec{v}_o = 50\vec{\imath} + 10\vec{\jmath} \text{ m/s}.$$

Note that the subscript o indicates the value at time $t = 0$ and not to be confused with that for the origin of the Newtonian frame of reference (*NFR*).

Similarly, the prime denotes the position vector measured from mass center G. That is, the symbols adopted in the present solution are consistent with those in Sections 6.4 and 6.5, unless it is stated otherwise.

Location of mass center,
Since G is the mass center, thus

$$m_A r'_A = m_B r'_B \quad \Rightarrow \quad \frac{r'_A}{m_B} = \frac{r'_B}{m_A}.$$

Adding $m_A r'_B$ to both sides of the first equation, one has

$$m_A r'_A + m_A r'_B = m_B r'_B + m_A r'_B$$

$$\Rightarrow \quad m_A \left(r'_A + r'_B \right) = r'_B \left(m_A + m_B \right)$$

$$\Rightarrow \quad \frac{r'_A + r'_B}{m_A + m_B} = \frac{r'_B}{m_A} \quad \Rightarrow \quad \frac{\ell}{m} = \frac{r'_B}{m_A}$$

$$\Rightarrow \quad r'_B = \frac{\ell}{m} m_A = \left(\frac{10}{180} \right) 100 = 5.5556 \text{ m}.$$

Therefore,

$$r'_A = AG = \ell - r'_B = 4.4444 \text{ m}.$$

At initial conditions
Linear momentum,

$$\vec{L}_o = m\vec{v}_o = (180 \text{ kg}) \left(50\vec{\imath} + 10\vec{\jmath} \text{ m/s} \right)$$
$$= 10^3 \left(9\vec{\imath} + 1.8\vec{\jmath} \right) \text{ kg.m/s}.$$

Angular momentum,
Angular momentum about mass center,

$$\left(\vec{H}_G \right)_o = \vec{r}'_A \times m_A \vec{v}'_A + \vec{r}'_B \times m_B \vec{v}'_B, \tag{6.39a}$$

where the subscript o, as noted earlier, denotes at $t = 0$.

Consider the first term on the rhs of Equation (6.39a). With reference to Figure 6.6c,

$$\vec{r}'_A = x_A \vec{\imath} + (y_A - \bar{y}_o) \vec{\jmath},$$

$$\vec{v}'_A = (r'_A \omega) \left[-\cos(90° - \beta) \, \vec{\imath} + \sin(90° - \beta) \, \vec{\jmath} \right]$$
$$= (r'_A \omega) \left(-\sin \beta \, \vec{\imath} + \cos \beta \, \vec{\jmath} \right).$$

Therefore,

$$\vec{r}'_A \times m_A \vec{v}'_A = \left[x_A \vec{\imath} + (y_A - \bar{y}_o) \, \vec{\jmath} \right] \times$$
$$m_A (r'_A \omega) \left(-\sin \beta \, \vec{\imath} + \cos \beta \, \vec{\jmath} \right)$$
$$= m_A (r'_A \omega) \left[x_A \cos \beta \, \vec{k} + (y_A - \bar{y}_o) \sin \beta \, \vec{k} \right]$$
$$= m_A (r'_A \omega) \left[x_A \cos \beta + (y_A - \bar{y}_o) \sin \beta \right] \vec{k}$$

in which

$$x_A \cos \beta + (y_A - \bar{y}_o) \sin \beta = r'_A.$$

Thus,

$$\vec{r}'_A \times m_A \vec{v}'_A = m_A \omega \left(r'_A \right)^2 \vec{k}. \tag{6.39b}$$

This equation simply states the fact that the direction of the angular momentum of satellite A about G is along $\vec{k}$ and the magnitude of the angular momentum is the product of r'_A and the linear momentum $m_A v'_A$ in which $v'_A = \omega r'_A$.

Similarly, the second term on the rhs of Equation (6.39a)

$$\vec{r}'_B \times m_B \vec{v}'_B = m_B \omega \left(r'_B \right)^2 \vec{k}. \tag{6.39c}$$

Substituting Equations (6.39b) and (6.39c) into (6.39a), one obtains

$$\left(\vec{H}_G \right)_o = m_A \omega \left(r'_A \right)^2 \vec{k} + m_B \omega \left(r'_B \right)^2 \vec{k}$$
$$= \omega \left[m_A \left(r'_A \right)^2 + m_B \left(r'_B \right)^2 \right] \vec{k}$$
$$= 20 \left[100 (4.4444)^2 + 80 (5.5556)^2 \right] \vec{k} \text{ kg.m}^2/\text{s}$$
$$= 8.8890 \left(10^4 \right) \vec{k} \text{ kg.m}^2/\text{s}.$$

Angular momentum about origin O,

$$\left(\vec{H}_O \right)_o = \vec{r} \times m \vec{v}_o + \left(\vec{H}_G \right)_o$$
$$= (\bar{y}_o) \, \vec{\jmath} \times \left(10^3 \right) \left(9 \vec{\imath} + 1.8 \vec{\jmath} \right)$$
$$+ 8.8890 \left(10^4 \right) \vec{k} \text{ kg.m}^2/\text{s}$$
$$\Rightarrow \left(\vec{H}_O \right)_o = - (10) \left(10^3 \right) (9) \, \vec{k} + (8.8890) \left(10^4 \right) \vec{k} \text{ kg.m}^2/\text{s}$$
$$\Rightarrow \left(\vec{H}_O \right)_o = -1.11 \left(10^3 \right) \vec{k} \text{ kg.m}^2/\text{s}.$$

Kinetic energy,

By Equation (6.34),

$$T_o = \frac{1}{2}m\bar{v}_o^2 + \frac{1}{2}m_A \left(v_A'\right)^2 + \frac{1}{2}m_B \left(v_B'\right)^2.$$

Therefore,

$$T_o = \frac{1}{2}(180)\left(50^2 + 10^2\right) + \frac{1}{2}(100)\left[(4.4444)(20)\right]^2$$
$$+ \frac{1}{2}(80)\left[(5.5556)(20)\right]^2 \text{ J}$$
$$\Rightarrow \quad T_o = 11.2289\left(10^5\right) \text{ J}.$$

(a) *Velocities of satellites*

Conservation of linear momentum,

$$\vec{L}_o = \vec{L}.$$

With reference to Figure 6.6a and this equation, one has

$$10^3\left(9\vec{\imath} + 1.8\vec{\jmath}\right) \text{ kg.m/s} = m_A \vec{v}_A + m_B \vec{v}_B$$
$$= 100v_A\vec{\jmath} + 80\left[(v_B)_x \vec{\imath} + (v_B)_y \vec{\jmath}\right].$$

Equating coefficients of $\vec{\imath}$,

$$10^3 (9) = 80\,(v_B)_x \quad \Rightarrow \quad (v_B)_x = 112.5 \text{ m/s}. \tag{6.39d}$$

Equating coefficients of $\vec{\jmath}$,

$$10^3 (1.8) = 100v_A + 80\,(v_B)_y$$
$$\Rightarrow \quad (v_B)_y = 22.5 - 1.25v_A. \tag{6.39e}$$

Conservation of energy,

$$T_o = T \quad \Rightarrow \quad 11.2289\left(10^5\right) = \frac{1}{2}m_A\,(v_A)^2 + \frac{1}{2}m_B\,(v_B)^2.$$

Substituting Equations (6.39d) and (6.39e) into the above equation,

$$11.2289\left(10^5\right) = \frac{1}{2}(100)\,(v_A)^2 + \frac{1}{2}(80)\left[(v_B)_x^2 + (v_B)_y^2\right]$$
$$\Rightarrow \quad 11.2289\left(10^5\right) = 50\,(v_A)^2 + 40\left[112.5^2 + (22.5 - 1.25v_A)^2\right]$$
$$\Rightarrow \quad (v_A)^2 - 20v_A - 5301 = 0.$$

This quadratic equation gives $v_A = 83.49$ m/s or -63.49 m/s.

The negative value is inadmissible since v_A is pointing upward. Thus,

$$\vec{v}_A = 83.49\vec{j} \text{ m/s.}$$

Substituting this into Equation (6.39e),

$$(v_B)_y = 22.5 - 1.25\,(83.49) \text{ m/s} = -81.86 \text{ m/s.}$$

Therefore, the velocity of satellite B is

$$\vec{v}_B = (v_B)_x \, \vec{i} + (v_B)_y \, \vec{j} = 112.5\vec{i} - 81.86\vec{j} \text{ m/s.}$$

(b) **Distance h_A from y-axis**

Applying the principle of conservation of angular momentum about O,

$$\left(\vec{H}_O\right)_o = \vec{H}_O$$

$$\Rightarrow \quad -1.11\,(10^3)\,\vec{k} \text{ kg.m}^2/\text{s} = h_A \vec{i} \times m_A \vec{v}_A + h_B \vec{i} \times m_B \vec{v}_B$$

$$= h_A \vec{i} \times (100)\,(83.49\vec{j}) + (80)\,\vec{i} \times (80)\,(112.5\vec{i} - 81.86\vec{j})$$

$$= h_A \left(8349\vec{k}\right) - \left(523{,}904\,\vec{k}\right) \text{ kg.m}^2/\text{s.}$$

This gives the required distance from y-axis to satellite A,

$$h_A = 62.6176 \text{ m.}$$

6.10 EXERCISES

6.1. A system of three particles A, B, and C, has the following masses and position vectors:

$$m_A = 1.0 \text{ kg}, \quad m_B = 2.0 \text{ kg}, \quad m_C = 5.0 \text{ kg},$$

$$\vec{r}_A = 3\vec{i} + 3\vec{j} + 6\vec{k} \text{ m}, \quad \vec{r}_B = 3\vec{i} + 4\vec{j} + 2\vec{k} \text{ m},$$

and

$$\vec{r}_C = -3\vec{i} - 3\vec{j} - 5\vec{k} \text{ m.}$$

The velocities of these particles are given as, $\vec{v}_A = 5\vec{i} + 4\vec{j} + 3\vec{k}$ m/s, $\vec{v}_B = 2\vec{i} + 2\vec{j} + 5\vec{k} \frac{m}{s}$, and $\vec{v}_C = 6\vec{i} + 5\vec{j} + 4\vec{k}$ m/s. Determine

(a) the position vector $\bar{\vec{r}}$ of the mass center G of the system,

(b) the linear momentum $m\bar{\vec{v}}$ of the system, and

(c) the angular momentum $\vec{H}_G$ of the system.

6.2. An object of 10 kg is falling vertically along y-axis, as shown in Figure 6.7. At point E it explodes into three pieces A, B, and C whose masses are, respectively 3 kg, 2 kg, and 5 kg. Immediately after the explosion the velocity of each piece is directed as indicated in the figure and the velocity of A has been detected to be 100 m/s. Find the velocity of the 10 kg object just before the explosion.

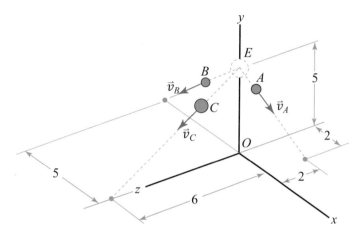

Figure 6.7: Explosion of an object (dimensions in m).

6.3. Three identical billiard balls are on a smooth horizontal table, as shown in Figure 6.8. Balls B and C are at rest and in contact while ball A moves with a velocity $\vec{v}_A$ to the right of the table. Given that the coefficient of restitution $e = 1.0$, determine the velocity of ball A immediately after impact if (a) its path is perfectly centered such that balls B and C are struck by ball A simultaneously and (b) ball A strikes ball C slightly before it strikes ball B.

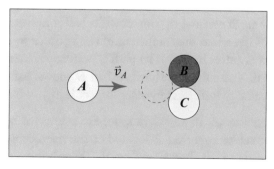

Figure 6.8: Three identical billiard balls on a smooth horizontal table.

6.4. Three identical billiard balls in a game of pool are on a smooth horizontal table. Ball A moves with a velocity magnitude $v_A = 20$ m/s, as indicated in Figure 6.9. If after collision the three balls move in the directions shown in the figure and the coefficient of restitution $e = 1.0$, determine the magnitudes of the velocities $\vec{v}\,'_A$, $\vec{v}\,'_B$, and $\vec{v}\,'_C$.

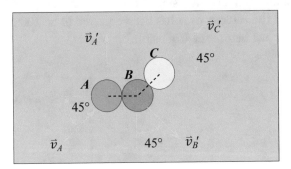

Figure 6.9: Three identical billiard balls in a game of pool.

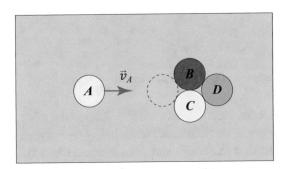

Figure 6.10: Four identical billiard balls on a smooth horizontal table.

6.5. Four identical billiard balls are on a smooth horizontal table, as shown in Figure 6.10. Balls B, C, and D are at rest and in contact while ball A moves with a velocity $\vec{v}_A$ to the right of the table. Given that the coefficient of restitution $e = 1$, determine the velocity of ball A immediately after impact if its path is perfectly centered such that balls B and C are struck by ball A simultaneously while ball D moves parallel to the longer side of the table and its velocity immediately after impact is $\vec{v}\,'_D = 0.5\vec{v}_A$.

6.6. Three identical satellites A, B, and C in space are connected by high strength inextensible and inelastic cables to a ring R located at the mass center of the three satellites, shown in Figure 6.11, such that $\ell_c = \ell\cos\theta$. The satellites are initially rotating in a plane about ring R which is at rest (stationary with respect to planet earth). The rotating speeds of the satellites are proportional to their distances from the ring R. When the original speeds of A and B are $v_A = v_B = v_o$ and the angle $\theta = 20°$, cable CR suddenly breaks. This causes satellite C to fly away. One is interested in the motion of the satellites A and B, and of the ring R after cables AR and BR have again become taut (That is, after C flies away and when cables AR and BR have become taut again R is the mass center of A and B). Determine (a) the speed of the ring R, and (b) the relative speed at which satellites A and B rotate about R.

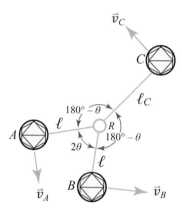

Figure 6.11: Three identical satellites in space.

C H A P T E R   7

Kinematics of Rigid Bodies

7.1 INTRODUCTION

In this chapter the kinematics of rigid bodies are presented. Section 7.1 is concerned with introduction to various important definitions and concepts in the kinematics of rigid bodies. The instantaneous center of rotation in plane motion is briefly introduced in Section 7.2. More details presentation and application in motion analysis may be found in textbooks on kinematics and dynamics of machinery, for example in [1–3]. Position vector in a rotating frame of reference is presented in Section 7.3. Rate of change of a vector with respect to a rotating frame of reference is dealt with in Section 7.4. The 3D motion of a point in a rigid body with respect to a rotating frame is considered in Section 7.5 in which three representative examples are included to illustrate the steps in the solution.

In a system of particles or rigid bodies, arguably the most fundamental concept is the *degrees-of-freedom* (dof) in spaces. By the dof of a system or rigid body one means the minimum *number of independent coordinates* that are required to completely describe the position of the system. It is independent of the coordinate system adopted in a particular situation. For example, if the system in a Cartesian coordinate system has 6 dof then it has the same number of dof in a spherical coordinate system. For a rigid body in 3D space, there are 3 translational dof, and 3 rotational dof, as shown in Figure 7.1. For a rigid body in a 2D space there are 2 translational dof and 1 rotational dof.

A rigid body may, in general, experiences either one or more of the various types of motion to be included in the following.

(a) *Translation* is the motion in which any straight line inside the body maintains the same direction, as shown in Figure 7.2a. Clearly, all particles in the rigid body move in parallel paths. However, if these paths are curved lines such motion is called *curvilinear translation*, as shown in Figure 7.2b.

(b) *Rotation about a fixed axis* is the motion in which the particles of the rigid body move in parallel planes along the circles centered on the same fixed axis, which is called the *axis of rotation*, as indicated in Figure 7.3.

(c) *General plane motion* is one which is neither a rotation nor a translation, as shown in Figure 7.4.

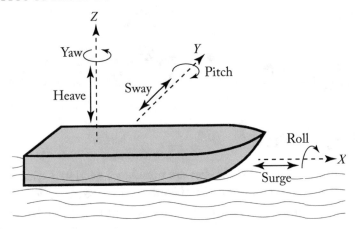

Figure 7.1: Ship floating in a sea.

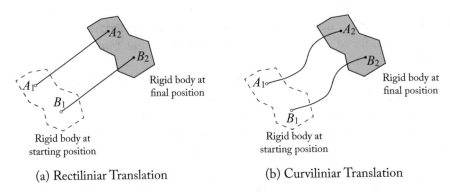

(a) Rectiliniar Translation   (b) Curviliniar Translation

Figure 7.2: (a) Rectilinear translation and (b) curvilinear translation.

(d) *Motion about a fixed point* is one that can better be understood through the Euler's theorem [4] which states that if a rigid body has one point fixed then any motion of that body can be reduced to a simple angular displacement about a single axis through the point. A classical example of this type of motion is the top on a rough floor, as shown in Figure 7.5.

It may be appropriate to note that if the component rotations in Euler's theorem are *finite* it is important that the *order* in which they are applied is kept. For infinitesimally small rotations the order is not important. In other words, for infinitesimally small rotations one can represent these rotations as vectors. On the hand, if these rotations are finite they cannot be represented as vectors. However, they can be represented as the so-called *pseudo-vectors* [5]. The latter quantities have important engineering applications and implications in more advanced dynamics analysis which is beyond the scope of the present book and therefore, will not be considered here.

Axis of Rotation

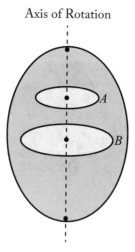

Figure 7.3: Rotation about a fixed axis.

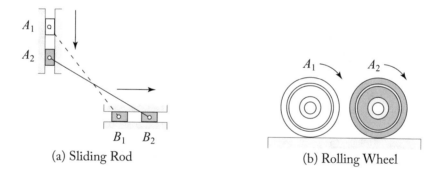

(a) Sliding Rod

(b) Rolling Wheel

Figure 7.4: General plane motion: (a) sliding rod in a mechanism and (b) rolling wheel.

(e) *General motion* is the type which does not belong to any of those presented in the foregoing. According to Charles' theorem [4] the most general displacement (motion) of a rigid body may be reduced to that of a translation, followed by a rotation.

Before leaving this section, the rotation of a rigid body about a fixed axis, and equations defining the rotation of a rigid body about a fixed axis should be introduced for completeness.

First, consider the rotation of the rigid body about a fixed axis, as shown in Figure 7.6 in which the fixed frame of reference $OXYZ$ is centered at the origin O, P is a point of the body and $\vec{r}$ is its position vector w.r.t. frame $OXYZ$, and B is called the *projection of P* on the fixed axis AA'. Note that the distance BP is constant such that when the rigid body rotates point P describes a circle whose radius is equal to $r \sin \varphi$, where φ is the angle between the position vector $\vec{r}$ and AA'. Note further that the position of P and the entire body is completely defined

Top in Motion

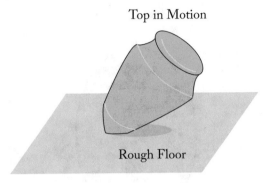

Rough Floor

Figure 7.5: Top on a rough floor.

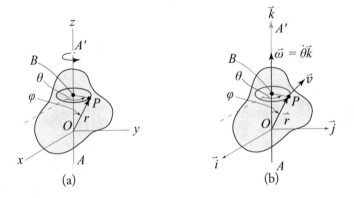

Figure 7.6: (a) Rotation of rigid body about a fixed axis and (b) velocity vector.

by the angle θ which is known as the *angular coordinate* of the body. The angle is expressed in radians and the right-hand screw rule is applied here.

   With reference to the definition for a particle in Chapter 2, one has the velocity of the point P as $\vec{v} = \frac{d\vec{r}}{dt}$ whose magnitude $v = \frac{ds}{dt} = r \sin \varphi \left(\frac{d\theta}{dt} \right)$. It should be noted that the angle θ depends on the position of P. However, the rate of change $\frac{d\theta}{dt} = \dot{\theta}$ is itself independent of P. With reference to Figure 7.6 or recall from vector analysis that the magnitude of the velocity given above is precisely the vector cross-product of two vectors, $\vec{\omega}$ and $\vec{r}$. That is,

$$\vec{v} = \frac{d\vec{r}}{dt} = \vec{\omega} \times \vec{r},$$

where the vector $\vec{\omega} = \omega\vec{k} = \dot{\theta}\vec{k}$. Thus, the acceleration of the point P of the rigid body is given by

$$\vec{a} = \frac{d\vec{v}}{dt} = \frac{d(\vec{\omega} \times \vec{r})}{dt} = \frac{d\vec{\omega}}{dt} \times \vec{r} + \vec{\omega} \times \frac{d\vec{r}}{dt} = \vec{\alpha} \times \vec{r} + \vec{\omega} \times \vec{v}.$$

Therefore,

$$\vec{a} = \vec{\alpha} \times \vec{r} + \vec{\omega} \times (\vec{\omega} \times \vec{r}).$$

Now, the equations defining the rotation of a rigid body about a fixed axis are introduced. Similar to those for rectilinear motion in Chapter 2, *uniform rotation* and *uniformly accelerated rotation* are frequently encountered in motion analysis. For the uniform rotation case, the angular velocity is constant so that the angular coordinate is given by

$$\theta = \theta_o + \omega t.$$

For the uniformly accelerated rotation case, one has the following equations which can similarly be derived as those presented in Chapter 2 for rectilinear motion:

$$\omega = \omega_o + \alpha t, \quad \theta = \theta_o + \omega_o t + \frac{1}{2}\alpha t^2, \quad \omega^2 = \omega_o^2 + 2\alpha(\theta - \theta_o).$$

7.2   INSTANTANEOUS CENTER OF ROTATION IN PLANE MOTION

The most *general motion of a rigid body in 2D space* (for example, the plane motion of a slab) may be treated *as sum of a translation and a rotation,* shown in Figure 7.7.

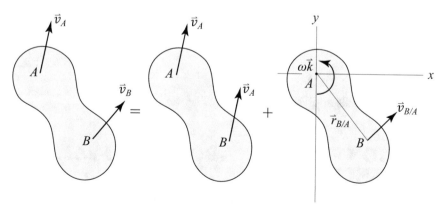

Figure 7.7: Plane motion of a slab in 2D space.

Consider the motion of the rigid body in a 2D space, as illustrated in Figure 7.7. It can be treated as a *translation* at point A and simultaneously a *rotation* about A. Thus, the velocity at point B of the 2D rigid body may be stated as

$$\vec{v}_B = \vec{v}_A + \vec{v}_{B/A}, \tag{7.1}$$

where $\vec{v}_{B/A} = \omega \vec{k} \times \vec{r}_{B/A}$ with $\vec{r}_{B/A}$ being the relative position vector of B relative to A, and $\vec{k}$ being the unit vector perpendicular to the plane of the rigid body and directed upward at point A (here, the right-hand screw has been assumed).

Another way to solve problems with velocities of points of a rigid body in the 2D space is by applying the *instantaneous (or simply instant) center of rotation* C of the rigid body, as shown in Figure 7.8. The instant center C is located by drawing lines perpendicular to the directions of the velocities at A and B, as shown in Figure 7.8. It may be appropriate to note that this *graphical method* is not applicable to the analysis of accelerations of rigid bodies because, in general, the acceleration at point C is not zero. For complicated systems that consist of many rigid bodies more elaborated steps are required in the locations of instant centers. They are not considered here since it is beyond the scope of the present book. However, for readers interested in such steps and their application they are referred to [3].

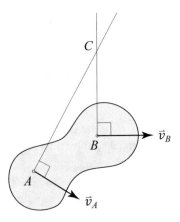

Figure 7.8: Motion analysis applying instantaneous center of rotation.

7.3 POSITION VECTOR IN A ROTATING FRAME OF REFERENCE

In the derivation of Equation (7.1), it was assumed the 2D rotation frame of reference (*RFR*) with its origin at A. In general, in a 3D space the position vector $\vec{r}_P$ connecting the origin of the *FFR*, *OXYZ* and the point of interest P may be expressed in terms of the unit vectors of a *RFR*, *Axyz* with its origin at point A and angular velocity $\vec{\Omega}$ as well as angular acceleration $\vec{\alpha}$, as shown in Figure 7.9.

With reference to Figure 7.9, the position of P is defined at any instant by the vector $\vec{r}_P$ in the *FFR*. Thus, one can write

$$\vec{r}_P = \vec{r}_A + \vec{r}_{P/A}. \qquad (7.2)$$

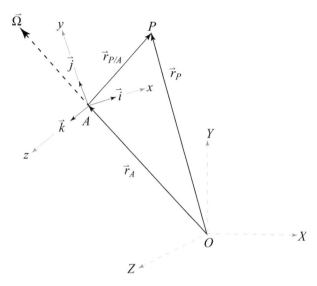

Figure 7.9: Position vector in a *FFR* and a *RFR*.

7.4   RATE OF CHANGE OF A VECTOR WITH RESPECT TO A ROTATING FRAME OF REFERENCE

Before one derives the equations for the velocity and acceleration at a point of interest by using Equation (7.2), consider the case in which the *FFR* and *RFR* have their origins at O, as shown in Figure 7.10.

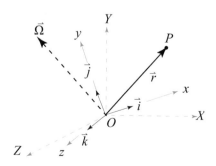

Figure 7.10: Origins O of *FFR* and *RFR*.

Let $\vec{\Omega}$ denotes the angular velocity of the frame $Oxyz$ or RFR at a given instant (with respect to $OXYZ$). Consider the position vector

$$\vec{r} = \vec{P} = P_x \vec{\imath} + P_y \vec{\jmath} + P_z \vec{k}, \qquad (7.3)$$

where P_x, P_y, and P_z are, respectively, the magnitudes of $\vec{P}$ resolving along the unit vectors $\vec{\imath}, \vec{\jmath}$, and $\vec{k}$ which are attached to the x, y, and z co-ordinates, respectively. In some textbooks, the latter co-ordinates are referred to as the body fixed co-ordinates (BFC). In this book the unit vectors $\vec{\imath}, \vec{\jmath}$, and $\vec{k}$ follow the right-hand screw rule, unless it is stated otherwise.

Differentiating Equation (7.3) with respect to time t and considering the unit vectors $(\vec{\imath}, \vec{\jmath}, \vec{k})$ as fixed (with respect to $Oxyz$), then the rate of change of $\vec{P}$ with respect to the RFR

$$\left(\frac{d\vec{P}}{dt}\right)_{Oxyz} = \frac{dP_x}{dt}\vec{\imath} + \frac{dP_y}{dt}\vec{\jmath} + \frac{dP_z}{dt}\vec{k}, \qquad (7.4)$$

where the subscript $Oxyz$ on the lhs of Equation (7.4) denotes the RFR.

To obtain the rate of change of $\vec{P}$ with respect to the FFR, $\left(\frac{d\vec{P}}{dt}\right)_{OXYZ}$ one must consider the unit vectors $(\vec{\imath}, \vec{\jmath}, \vec{k})$ as variables when operating Equation (7.3) because the RFR is rotating with respect to the FFR. In other words, the movement of RFR is time-dependent and thus,

$$\left(\frac{d\vec{P}}{dt}\right)_{OXYZ} = \frac{dP_x}{dt}\vec{\imath} + \frac{dP_y}{dt}\vec{\jmath} + \frac{dP_z}{dt}\vec{k} + P_x\frac{d\vec{\imath}}{dt} + P_y\frac{d\vec{\jmath}}{dt} + P_z\frac{d\vec{k}}{dt}. \qquad (7.5)$$

Applying Equation (7.4) to (7.5), the latter becomes

$$\left(\frac{d\vec{P}}{dt}\right)_{OXYZ} = \left(\frac{d\vec{P}}{dt}\right)_{Oxyz} + P_x\frac{d\vec{\imath}}{dt} + P_y\frac{d\vec{\jmath}}{dt} + P_z\frac{d\vec{k}}{dt}$$

or in more concise form

$$\left(\frac{d\vec{P}}{dt}\right)_{OXYZ} = \left(\frac{d\vec{P}}{dt}\right)_{Oxyz} + \vec{\Omega} \times \vec{P} \qquad (7.6)$$

where the second term on the rhs of Equation (7.6),

$$\vec{\Omega} \times \vec{P} = P_x\frac{d\vec{\imath}}{dt} + P_y\frac{d\vec{\jmath}}{dt} + P_z\frac{d\vec{k}}{dt}$$

because, for example, $\frac{d\vec{\imath}}{dt} = \vec{\Omega} \times \vec{\imath}$, which is the derivative of a unit position vector with respect to time t, is the velocity. In other words,

$$P_x\frac{d\vec{\imath}}{dt} + P_y\frac{d\vec{\jmath}}{dt} + P_z\frac{d\vec{k}}{dt} = P_x\vec{\Omega} \times \vec{\imath} + P_y\vec{\Omega} \times \vec{\jmath} + P_z\vec{\Omega} \times \vec{k}.$$

By factoring the common term $\vec{\Omega} \times$ from the rhs of the above equation, it becomes

$$P_x \frac{d\vec{\imath}}{dt} + P_y \frac{d\vec{\jmath}}{dt} + P_z \frac{d\vec{k}}{dt} = \vec{\Omega} \times \left(P_x \vec{\imath} + P_y \vec{\jmath} + P_z \vec{k} \right).$$

7.5 THREE-DIMENSIONAL MOTION OF A POINT IN A RIGID BODY WITH RESPECT TO A ROTATING FRAME

With reference to Figure 7.10, the absolute velocity $\vec{v}_P$ of the point is defined as the rate of change of $\vec{r}$ with respect to the *FFR* . Therefore, by making use of Equation (7.6), one has

$$\vec{v}_P = \left(\frac{d\vec{r}}{dt} \right)_{OXYZ} = \left(\frac{d\vec{r}}{dt} \right)_{Oxyz} + \vec{\Omega} \times \vec{r}, \tag{7.7}$$

where $\left(\frac{d\vec{r}}{dt} \right)_{Oxyz}$ is the velocity of the particle P relative to the *RFR*.

The absolute acceleration $\vec{a}_P$ of the particle P is defined as the rate of change of $\vec{v}_P$ with respect to the *FFR*. Thus, operating on Equation (7.7), one has

$$\vec{a}_P = \left(\frac{d\vec{v}_P}{dt} \right)_{OXYZ} = \left[\frac{d}{dt} \left(\frac{d\vec{r}}{dt} \right)_{Oxyz} \right]_{OXYZ}$$
$$+ \frac{d\vec{\Omega}}{dt} \times \vec{r} + \vec{\Omega} \times \left(\frac{d\vec{r}}{dt} \right)_{OXYZ}. \tag{7.8}$$

Applying Equation (7.6) to the first term on the rhs of the above equation, one has

$$\left[\frac{d}{dt} \left(\frac{d\vec{r}}{dt} \right)_{Oxyz} \right]_{OXYZ} = \left(\frac{d^2\vec{r}}{dt^2} \right)_{Oxyz} + \vec{\Omega} \times \left(\frac{d\vec{r}}{dt} \right)_{Oxyz}.$$

In addition, by making use of Equation (7.7) for part of the third term on the rhs of Equation (7.8), that is, $\left(\frac{d\vec{r}}{dt} \right)_{OXYZ} = \left(\frac{d\vec{r}}{dt} \right)_{Oxyz} + \vec{\Omega} \times \vec{r}$, so that the acceleration at point P becomes

$$\vec{a}_P = \left(\frac{d\vec{v}_P}{dt} \right)_{OXYZ}$$
$$= \left(\frac{d^2\vec{r}}{dt^2} \right)_{Oxyz} + \vec{\Omega} \times \left(\frac{d\vec{r}}{dt} \right)_{Oxyz}$$
$$+ \frac{d\vec{\Omega}}{dt} \times \vec{r} + \vec{\Omega} \times \left[\left(\frac{d\vec{r}}{dt} \right)_{Oxyz} + \vec{\Omega} \times \vec{r} \right].$$

Collecting common terms and re-arranging, one obtains

$$\vec{a}_P = \left(\frac{d^2\vec{r}}{dt^2}\right)_{Oxyz} + \vec{\alpha} \times \vec{r} + \vec{\Omega} \times \left(\vec{\Omega} \times \vec{r}\right) + 2\vec{\Omega} \times \left(\frac{d\vec{r}}{dt}\right)_{Oxyz} \tag{7.9}$$

in which the angular acceleration $\vec{\alpha} = \frac{d\vec{\Omega}}{dt}$ has been used.

Returning to the more general case as shown in Figure 7.9 in which the origins of the *FFR* and *RFR* are not at the same point, and by making use of Equations (7.2) and (7.7), one can show that

$$\vec{v}_P = \left(\frac{d\vec{r}_P}{dt}\right)_{OXYZ} = \vec{v}_A + \left(\frac{d\vec{r}_{P/A}}{dt}\right)_{Oxyz} + \vec{\Omega} \times \vec{r}_{P/A}, \tag{7.10}$$

where $\vec{v}_A$ is the velocity at the origin A of the *RFR*, $Axyz$.

Similarly, by making use of Equations (7.2) and (7.9), the corresponding acceleration becomes

$$\vec{a}_P = \left(\frac{d\vec{v}_P}{dt}\right)_{OXYZ} = \vec{a}_A + \left(\frac{d^2\vec{r}_{P/A}}{dt^2}\right)_{Oxyz}$$

$$+ \vec{\alpha} \times \vec{r}_{P/A} + \vec{\Omega} \times \left(\vec{\Omega} \times \vec{r}_{P/A}\right) + 2\vec{\Omega} \times \left(\frac{d\vec{r}_{P/A}}{dt}\right)_{Oxyz}, \tag{7.11}$$

where $\vec{a}_A$ is the acceleration at the origin A of the *RFR* and the last term on the rhs, $2\vec{\Omega} \times \left(\frac{d\vec{r}_{P/A}}{dt}\right)_{Oxyz}$ is called the *Coriolis* (after the French mathematician De Coriolis, 1792–1843) or *complementary acceleration*. It should be noted that in general every term in Equation (7.11) has two components, the normal and tangential components, for example.

In many problems, Equation (7.10) can be rewritten as

$$\vec{v}_P = \vec{v}_{P'} + \vec{v}_{P/F}, \tag{7.12}$$

where $\vec{v}_P$ = absolute instantaneous velocity of point P,
  $\vec{v}_{P'}$ = instantaneous velocity of point P' of *RFR* coinciding with P, and
  $\vec{v}_{P/F}$ = instantaneous velocity of point P relative to the *RFR*.
In what follows the adjective "instantaneous" will be disregarded for brevity.

Similarly, Equation (7.11) can be rewritten as

$$\vec{a}_P = \vec{a}_{P'} + \vec{a}_{P/F} + \vec{a}_c, \tag{7.13}$$

where $\vec{a}_P$ = absolute acceleration of point P,
  $\vec{v}_{P'}$ = acceleration of point P' of the *RFR* coinciding with P,
  $\vec{a}_{P/F}$ = acceleration of point P relative to the *RFR*, and

$\vec{a}_c = 2\vec{\Omega} \times \left(\frac{d\vec{r}_{P/A}}{dt}\right)_{Oxyz}$ = complementary or Coriolis acceleration.

The approach presented in the foregoing is generally known as the *method of moving frame of reference* or *unit vector method* since every term in Equations (7.10) and (7.11) is expressed in terms of the unit vectors of the *RFR*. The application of Equations (7.12) and (7.13) is introduced in the next two examples while the use of Equations (7.10) and (7.11) is illustrated in the third example.

Example 7.1
A circular disk of uniform thickness and horizontal shafts of uniform cross-section attached to both ends of the diameter of the disk. The horizontal shafts being held by bearings O and H, as shown in Figure 7.11a, rotate at a constant angular velocity $\omega_1 = 1.0$ rad/s. Suppose the position of the disk shown in the figure is in the XY-plane (Z-axis is perpendicular to this plane) and point B of the strap AB travels as indicated at a constant relative speed $v = 2.0$ m/s. If the length of the strap $AB = 200$ mm, find (a) the velocity at B and (b) acceleration at B.

Solution:
Let $OXYZ$ and $Axyz$ be the *FFR* and *RFR*, respectively, with A being the origin of the *RFR* which is assumed to attach to the rotating disk.

Position or Geometry
With reference to the given figure, the relative position vector

$$\vec{r}_{B/A} = (0.2 \text{ m}) \left(\cos 20° \vec{\imath} - \sin 20° \vec{\jmath}\right)$$
$$= (0.2 \text{ m}) \left(0.9397\vec{\imath} - 0.3420\vec{\jmath}\right)$$
$$= 0.1879\vec{\imath} - 0.0684\vec{\jmath} \text{ m.}$$

(a) *Velocity analysis*

In this part of the analysis Equation (7.12) is employed,

$$\vec{v}_P = \vec{v}_{P'} + \vec{v}_{P/F}$$

in which the angular velocity of the *RFR* is

$$\vec{\Omega} = \omega_1 \vec{\jmath} = 1.0\vec{\jmath} \text{ rad/s}$$

and P is B in the present problem. Thus, velocity of point B' of *RFR* coinciding with B,

$$\vec{v}_{B'} = \vec{\Omega} \times \vec{r}_{B/A} = 1.0\vec{\jmath} \times \left(0.1879\vec{\imath} - 0.0684\vec{\jmath}\right) \text{ m/s}$$
$$= -0.1879\vec{k} \text{ m/s.}$$

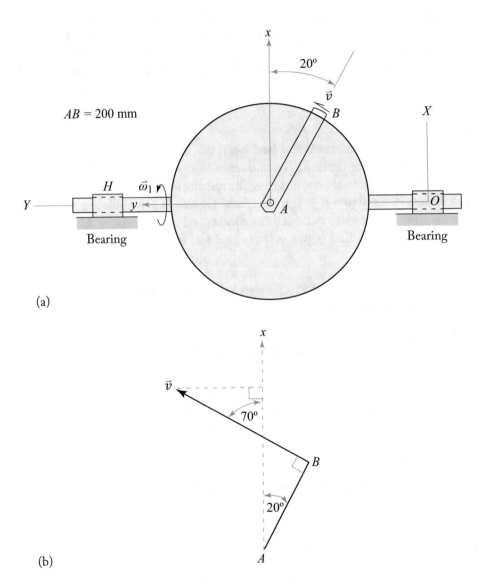

(a)

(b)

Figure 7.11: Circular disk rotating about Y-axis (Z and z are perpendicular to the XY-plane and not shown in the figure): (a) circular disk with *FFR* and *RFR* and (b) direction of velocity at *B* relative to *RFR*.

With reference to Figure 7.11b, velocity of point B relative to the *RFR* is

$$\vec{v}_{B/F} = v\left(\cos 70° \vec{\imath} + \sin 70° \vec{\jmath}\right) = 2\left(0.3420\vec{\imath} + 0.9397\vec{\jmath}\right) \text{ m/s}$$
$$= 0.6840\vec{\imath} + 1.8794\vec{\jmath} \text{ m/s}.$$

Therefore, the velocity at point B is

$$\vec{v}_B = \vec{v}_{B'} + \vec{v}_{B/F} = -0.1879\vec{k} + 0.6840\vec{\imath} + 1.8794\vec{\jmath} \text{ m/s}$$
$$= 0.6840\vec{\imath} + 1.8794\vec{\jmath} - 0.1879\vec{k} \text{ m/s}.$$

(b) *Acceleration analysis*

Applying Equation (7.13),

$$\vec{a}_P = \vec{a}_{P'} + \vec{a}_{P/F} + \vec{a}_c$$

in which the acceleration of point B' of *RFR* coinciding with B is

$$\vec{a}_{B'} = \vec{\alpha} \times \vec{r}_{B/A} + \vec{\Omega} \times \left(\vec{\Omega} \times \vec{r}_{B/A}\right) = \vec{\Omega} \times \left(\vec{\Omega} \times \vec{r}_{B/A}\right)$$

since $\vec{\Omega} = \omega_1 \vec{\jmath}$ is constant and therefore $\vec{\alpha} = 0$. This gives

$$\vec{a}_{B'} = \vec{\Omega} \times \left(\vec{\Omega} \times \vec{r}_{B/A}\right) = \omega_1 \vec{\jmath} \times \left[\omega_1 \vec{\jmath} \times \left(0.1879\vec{\imath} - 0.0684\vec{\jmath}\right)\right]$$
$$= 1.0\vec{\jmath} \times \left[1.0\vec{\jmath} \times \left(0.1879\vec{\imath} - 0.0684\vec{\jmath}\right)\right] \text{ m/s}^2$$
$$= -1.0\vec{\jmath} \times 0.1879\vec{k} \text{ m/s}^2 = -0.1879\vec{\imath} \text{ m/s}^2.$$

Acceleration of point B relative to the *RFR*,

$$\vec{a}_{B/F} = \left(\vec{a}_{B/F}\right)_n + \left(\vec{a}_{B/F}\right)_t$$

in which the tangential component is zero due to the fact that $\vec{v}_{B/F}$ is constant and therefore its time derivative is zero while the normal component pointing from B toward the center of curvature A is

$$\vec{a}_{B/F} = \left(\vec{a}_{B/F}\right)_n = \frac{v^2}{(AB)}\left(-\cos 20° \vec{\imath} + \sin 20° \vec{\jmath}\right)$$
$$= \frac{2.0^2}{(0.2)}\left(-0.9397\vec{\imath} + 0.3420\vec{\jmath}\right) \text{ m/s}^2$$
$$= -18.7940\vec{\imath} + 6.8400\vec{\jmath} \text{ m/s}^2.$$

The Coriolis acceleration is

$$\vec{a}_c = 2\vec{\Omega} \times \left(\frac{d\vec{r}_{B/A}}{dt}\right)_{Axyz} = 2\left(\omega_1\vec{j} \times \vec{v}_{B/F}\right)$$

$$= 2.0\vec{j} \times \left(0.6840\vec{i} + 1.8794\vec{j}\right) = -1.3680\vec{k} \text{ m/s}^2.$$

Summing all the terms, one obtains the acceleration at point B as

$$\vec{a}_B = \vec{a}_{B'} + \vec{a}_{B/F} + \vec{a}_c$$

$$= -0.1879\vec{i} - 18.7940\vec{i} + 6.8400\vec{j} - 1.3680\vec{k} \text{ m/s}^2$$

$$= -18.9819\vec{i} + 6.8400\vec{j} - 1.3680\vec{k} \text{ m/s}^2.$$

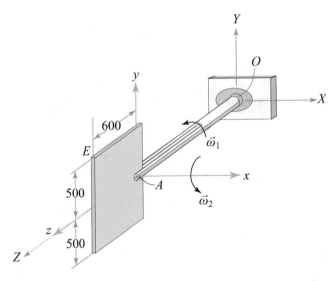

Figure 7.12: Rectangular plate rotating about Z-axis (dimensions in mm).

Example 7.2

A rectangular plate of uniform thickness rotates at a constant angular speed $\omega_2 = 10$ rad/s with respect to the horizontal arm AO, as shown in Figure 7.12. The arm AO rotates at constant angular speed $\omega_1 = 20$ rad/s about the fixed Z-axis (which coincides with the z-axis of the RFR, as shown in the figure) of the *FFR OXYZ* (which is known as the Newtonian or inertial frame of reference). For the position shown, determine the (a) velocity and (b) acceleration of the point E.

Solution:

Let the *RFR* be attached to the rectangular plate as shown in Figure 7.12. With reference to the latter figure and expressing all quantities in terms of unit vectors of the *RFR*, the relative position vector becomes

$$\vec{r}_{E/A} = 0.500\vec{j} + 0.600\vec{k} \text{ m.}$$

(a) *Velocity analysis*

In this part of the analysis Equation (7.12) is employed,

$$\vec{v}_P = \vec{v}_{P'} + \vec{v}_{P/F}$$

in which the angular velocity of the *RFR* is

$$\vec{\Omega} = \omega_1\vec{k} = 20.0\vec{k} \text{ rad/s}$$

and the point P is E in the present problem. Thus, velocity of point E' of *RFR* coinciding with E is

$$\vec{v}_{E'} = \vec{\Omega} \times \vec{r}_{E/A} = 20.0\vec{k} \times \left(0.500\vec{j} + 0.600\vec{k}\right) \text{ m/s}$$
$$= -10.0\vec{i} \text{ m/s.}$$

Velocity of point E relative to the *RFR*,

$$\vec{v}_{E/F} = \vec{\omega}_2 \times \vec{r}_{E/A} = 10.0\vec{i} \times \left(0.500\vec{j} + 0.600\vec{k}\right) \text{ m/s}$$
$$= -6.000\vec{j} + 5.000\vec{k} \text{ m/s.}$$

Therefore, the velocity at point E is

$$\vec{v}_E = \vec{v}_{E'} + \vec{v}_{E/F} = -10.0\vec{i} - 6.000\vec{j} + 5.000\vec{k} \text{ m/s.}$$

(b) *Acceleration analysis*

Applying Equation (7.13),

$$\vec{a}_P = \vec{a}_{P'} + \vec{a}_{P/F} + \vec{a}_c$$

in which the acceleration of point E' of *RFR* coinciding with E is

$$\vec{a}_{E'} = \vec{\alpha} \times \vec{r}_{E/A} + \vec{\Omega} \times \left(\vec{\Omega} \times \vec{r}_{E/A}\right) = \vec{\Omega} \times \left(\vec{\Omega} \times \vec{r}_{E/A}\right)$$

since $\vec{\Omega} = \omega_1\vec{k}$ is constant and therefore $\vec{\alpha} = 0$. This gives

$$\vec{a}_{E'} = \vec{\Omega} \times \left(\vec{\Omega} \times \vec{r}_{E/A}\right) = \omega_1\vec{k} \times \left[\omega_1\vec{k} \times \left(0.500\vec{j} + 0.600\vec{k}\right)\right]$$
$$= 20.0\vec{k} \times \left(-10.0\vec{i}\right) \text{ m/s}^2$$
$$= -200.0\vec{j} \text{ m/s}^2.$$

Acceleration of point E relative to the *RFR*,

$$\vec{a}_{E/F} = \vec{\alpha}_2 \times \vec{r}_{E/A} + \vec{\omega}_2 \times (\vec{\omega}_2 \times \vec{r}_{E/A}) = \vec{\omega}_2 \times (\vec{\omega}_2 \times \vec{r}_{E/A})$$

in which the tangential component is zero due to the fact that $\vec{\omega}_2$ is constant and therefore $\vec{\alpha}_2$ is zero. Therefore,

$$\vec{a}_{E/F} = \vec{\omega}_2 \times (\vec{\omega}_2 \times \vec{r}_{E/A}) = 10.0\vec{i} \times \left[10.0\vec{i} \times \left(0.500\vec{j} + 0.600\vec{k} \right) \right] \text{ m/s}^2$$

$$= 10.0\vec{i} \times \left(-6.000\vec{j} + 5.000\vec{k} \right) \text{ m/s}^2$$

$$= -50.0\vec{j} - 60.0\vec{k} \text{ m/s}^2.$$

The Coriolis acceleration is

$$\vec{a}_c = 2\vec{\Omega} \times \left(\frac{d\vec{r}_{E/A}}{dt} \right)_{Axyz} = 2 \left(\omega_1 \vec{k} \times \vec{v}_{E/F} \right)$$

$$= 40.0\vec{k} \times \left(-6.000\vec{j} + 5.000\vec{k} \right) = 240.0\vec{i} \text{ m/s}^2.$$

Summing all the terms, one obtains the acceleration at point E as

$$\vec{a}_E = \vec{a}_{E'} + \vec{a}_{E/F} + \vec{a}_c$$

$$= -200.0\vec{j} - 50.0\vec{j} - 60.0\vec{k} + 240.0\vec{i} \text{ m/s}^2$$

$$= 240.0\vec{i} - 250.0\vec{j} - 60.0\vec{k} \text{ m/s}^2.$$

Example 7.3

The cam mechanism shown in Figure 7.13a [3] has a constant angular velocity $\omega_2 = 2.0$ rad/s clockwise. The diameter of the roller (rigid body or link 3) of the follower is 12.7 mm. The dimensions in this figure are in mm. By applying the method of unit vectors, determine the velocity and acceleration at point A of the follower (more precisely, follower shaft, rigid body, or link 4) for the phase shown. Given that the angle O_2AB is 64.47° (note that O_2B is parallel to the y-axis of the *RFR*).

Solution:

The solution of this problem consists of three parts. They are the (a) position analysis, (b) velocity analysis, and (c) acceleration analysis.

Let O_2 be the origin of the *FFR* and A be the origin of the *RFR* with x-axis parallel to the flat face of the cam as shown in the sketch in Figure 7.13b. Imagine A is attached to link 2 then A_2 denotes the point A associated with link 2. The point A_4 of interest is attached to the shaft of the follower (that is, A_4 is the point A associated with link 4). The points A_2 and A_4 are the so-called *co-incident points*.

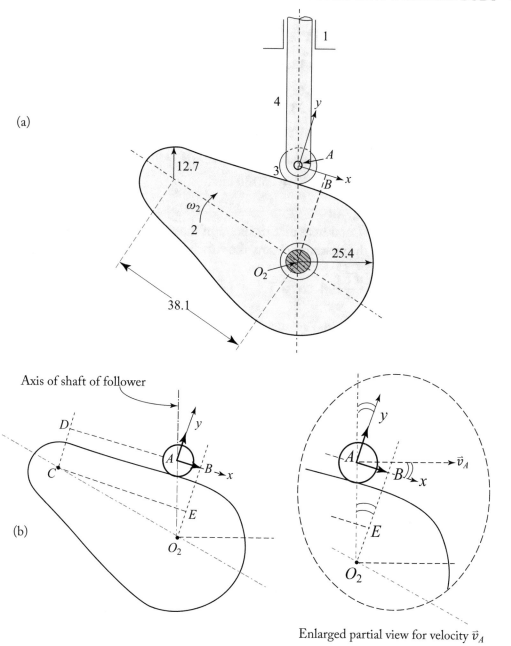

Figure 7.13: (a) Cam mechanism (dimensions in mm) and (b) Sketches of cam mechanism for position and velocity analyses.

(a) *Position analysis*

With reference to Figure 7.13b:

$$CD = 12.7 + 12.7/2 \text{ mm} = 19.05 \text{ mm.}$$
$$O_2B = 25.4 + 12.7/2 \text{ mm} = 31.75 \text{ mm.}$$
$$O_2C = 38.1 \text{ mm.}$$
$$O_2E = O_2B - CD = 12.7 \text{ mm.}$$

Since $\angle O_2AB = 64.47°$, given, and therefore $\angle AO_2B = 90° - 64.47° = 25.53°$. Thus, $O_2A = O_2B/(\cos \angle AO_2B) = 35.186 \text{ mm.}$

(b) *Velocity analysis*

From Equation (7.10), and note that the point of interest P there is A_4 in this problem and A there is A_2 in the present problem, the velocity of interest is

$$\vec{v}_P = \vec{v}_{A_4} = \left(\frac{d\vec{r}_{A_4}}{dt}\right)_{OXYZ} = \vec{v}_{A_2} + \left(\frac{d\vec{r}_{A_4/A_2}}{dt}\right)_{Oxyz} + \vec{\Omega} \times \vec{r}_{A_4/A_2}. \qquad (7.14a)$$

The terms in Equation (7.14a) are now identified.

Since the follower can only move along the axis of its shaft (because it is constrained by the bearing not shown in the figure) and with reference to Figures 7.13a,b,

$$\vec{v}_{A_4} = v_{A_4}\left(-\sin 25.53° \vec{i} + \cos 25.53° \vec{j}\right)$$
$$= v_{A_4}\left(-0.4310\vec{i} + 0.9024\vec{j}\right).$$

The magnitude of the velocity v_{A_4} is unknown while the direction is along the axis of its shaft as indicated in the foregoing.

The velocity at A is perpendicular to O_2A (recall, the velocity at a point is always perpendicular to the position vector) and its magnitude is known since the length O_2A and ω_2 are determined and given, respectively. Since A_2 is associated with link 2, thus line O_2A can be considered as rotating with the angular velocity of link 2, $\vec{\omega}_2$ which is $\vec{\Omega}$ in Equation (7.14a). Hence,

$$\vec{v}_A = \vec{v}_{A_2} = \omega_2 (O_2A)\left(\cos 25.53° \vec{i} + \sin 25.53° \vec{j}\right)$$
$$= 2 (35.186)\left(\cos 25.53° \vec{i} + \sin 25.53° \vec{j}\right) \text{ mm/s}$$
$$= 63.50\vec{i} + 30.33\vec{j} \text{ mm/s.}$$

The relative velocity of P with respect to the origin of the *RFR* referring to the *RFR* is

$$\left(\frac{d\vec{r}_{A_4/A_2}}{dt}\right)_{Oxyz} = v\vec{i}.$$

That is, in general, the relative velocity at P or A_4 with respect to A_2 is not zero as A_4 and A_2 are associated with different links (rigid bodies). One may wonder why the last equation holds true. The implicit assumption in this problem is that the roller of the follower does not separate from the surface of the cam and it does not penetrate into the cam. In other words, the only relative motion possible is along the surface of the cam. However, the last term on the rhs of Equation (7.14a),

$$\vec{\Omega} \times \vec{r}_{A_4/A_2} = 0, \quad \text{since} \quad \vec{r}_{A_4/A_2} = 0.$$

Substituting all the terms into Equation (7.14a), one has

$$v_P \left(-0.4310\vec{i} + 0.9024\vec{j} \right) = 63.50\vec{i} + 30.33\vec{j} + v\vec{i}. \tag{7.14b}$$

Equating coefficients of terms associated with $\vec{j}$ results

$$v_P (0.9024) = 30.33,$$

giving $v_P = 33.61$ mm/s.

Substituting into Equation (7.14b) and equating coefficients of terms associated with $\vec{i}$, one finds $v = -77.986$ mm/s. Therefore, the required velocity is

$$\vec{v}_P = \vec{v}_{A_4} = -14.486\vec{i} + 30.33\vec{j} \text{ mm/s}.$$

(c) *Acceleration analysis*

The acceleration from Equation (7.11) is

$$\vec{a}_{A_4} = \vec{a}_{A_2} + \left(\frac{d^2\vec{r}_{A_4/A_2}}{dt^2} \right)_{Axyz}$$

$$+ \vec{\alpha} \times \vec{r}_{A_4/A_2} + \vec{\Omega} \times \left(\vec{\Omega} \times \vec{r}_{A_4/A_2} \right) + 2\vec{\Omega} \times \left(\frac{d\vec{r}_{A_4/A_2}}{dt} \right)_{Axyz}.$$

Since $\vec{r}_{P/A} = \vec{r}_{A_4/A_2} = 0$, therefore it reduces to

$$\vec{a}_{A_4} = \vec{a}_{A_2} + \left(\frac{d^2\vec{r}_{A_4/A_2}}{dt^2} \right)_{Axyz} + 2\vec{\Omega} \times \left(\frac{d\vec{r}_{A_4/A_2}}{dt} \right)_{Axyz}. \tag{7.14c}$$

Consider first the lhs term,

$$\vec{a}_{A_4} = \left(\vec{a}_{A_4} \right)_n + \left(\vec{a}_{A_4} \right)_t$$

in which the normal component $\left(\vec{a}_{A_4} \right)_n = 0$ because the follower is supposed to move along the axis of the shaft only. Otherwise, the follower would not move because the non-zero normal component (perpendicular to the axis of the shaft of the follower) of the acceleration at A_4 will result in a couple that will jam the follower. Thus, the tangential component (along the axis of the shaft of the follower; also parallel to the velocity at A_4) is

$$
\begin{aligned}
(\vec{a}_{A_4})_t &= (a_{A_4})_t \left(-\sin 25.53° \vec{i} + \cos 25.53° \vec{j}\right) \\
&= (a_{A_4})_t \left(-0.4310 \vec{i} + 0.9024 \vec{j}\right).
\end{aligned}
$$

The first term on the rhs of Equation (7.14c) is

$$
\begin{aligned}
\vec{a}_{A_2} &= \omega_2^2 (O_2 A) \left(\sin 25.53° \vec{i} - \cos 25.53° \vec{j}\right) \\
&= 60.66 \vec{i} - 127.0 \vec{j} \text{ mm/s}^2.
\end{aligned}
$$

This is the normal component and it is along the axis of the shaft of the follower. The direction is from A_2 toward O_2 (recall, the normal component of the acceleration at a point is always directing toward the center of curvature in a circular motion). The tangential component associated with the angular acceleration is zero since the given angular velocity $\vec{\omega}_2$ of the *RFR* is constant. Physically, if this tangential component is not zero it would means a force along this direction will result a moment jamming the shaft of the follower. In order to avoid jamming the shaft of the follower the tangential component of this acceleration should be zero and therefore the angular velocity $\vec{\omega}_2$ has to be constant.

The second term on the rhs of Equation (7.14c) is

$$
\begin{aligned}
\left(\frac{d^2 \vec{r}_{A_4/A_2}}{dt^2}\right)_{Axyz} &= \left(\frac{d^2 \vec{r}_{A_4/A_2}}{dt^2}\right)^t_{Axyz} + \left(\frac{d^2 \vec{r}_{A_4/A_2}}{dt^2}\right)^n_{Axyz} \\
&= \left(\frac{d^2 \vec{r}_{P/A}}{dt^2}\right)^t_{Axyz} = a \vec{i},
\end{aligned}
$$

where the superscript t and n denote, respectively, the tangential and normal components of the acceleration. The magnitude of the acceleration a is unknown. Since only the relative acceleration parallel (that is, tangential) to the direction of the flat surface of the cam is possible (that is, the normal component is zero; otherwise, the roller of the follower will leave or lose contact with the surface of the cam or penetrate into the surface which is impossible as all the links are assumed to be rigid bodies).

The third term on the rhs of Equation (7.14c) is

$$
2 \vec{\Omega} \times \left(\frac{d \vec{r}_{A_4/A_2}}{dt}\right)_{Axyz} = 2\left(-\omega_2 \vec{k}\right) \times v \vec{i} = 2\left(-\omega_2 \vec{k}\right) \times (-77.986) \vec{i}
$$

$$
= 311.944 \vec{j} \text{ mm/s}^2.
$$

Substituting all the above terms into Equation (7.14c), one obtains

$$\left(a_{A_4}\right)_t \left(-0.4310\vec{i} + 0.9024\vec{j}\right) = 60.66\vec{i} - 127.0\vec{j} + a\vec{i} + 311.944\vec{j}.$$

Equating coefficients associated with unit vector $\vec{j}$, one has

$$(a_P)_t \, (0.9024) = -127.0 + 311.944$$

which gives

$$\left(a_{A_4}\right)_t = a_{A_4} = 204.947 \text{ mm/s}^2.$$

Thus, the required acceleration of the follower at A_4 is

$$\vec{a}_P = \vec{a}_{A_4} = -88.33\vec{i} + 184.94\vec{j} \text{ mm/s}^2.$$

Remarks:

In this example, application of the unit vector method based on Equations (7.10) and (7.11) has been made. In applying these equations to the example, the terms on the lhs are referenced to the *RFR*. However, in the derivation of Equations (7.10) and (7.11) the terms on the lhs are referenced to the *FFR*. This, naturally, leads to the question of the validity of the solution. To show that whether the terms on the lhs of the equations referenced to the *RFR* or *FFR* are equal the following simple illustration is in order.

Consider the desired velocity $\vec{v}_{A_4}$ of the example in terms of the unit vectors $\vec{I}, \vec{J}$, and $\vec{K}$ of the *FFR*. Note that the origin of the *FFR* is at O_2, X-axis is parallel to $\vec{v}_{A_2}$, Y-axis is along the axis of the shaft of follower, and Z-axis is perpendicular to the XY-plane. Thus,

$$\vec{v}_{A_4} = v_{A_4}\vec{J}.$$

With reference to Figure 7.13b, the unit vector $\vec{J}$ of the *FFR* is related to the unit vectors of the *RFR* by

$$\vec{J} = -\sin 25.53°\vec{i} + \cos 25.53°\vec{j}.$$

Therefore, the desired velocity

$$\vec{v}_{A_4} = v_{A_4}\left(-\sin 25.53°\vec{i} + \cos 25.53°\vec{j}\right).$$

This equation for $\vec{v}_{A_4}$ is identical to that referenced to the *RFR*.

7.6   EXERCISES

7.1. A circular disk of uniform thickness and horizontal shafts of uniform cross-section attached to both ends of the diameter of the disk. The horizontal shafts being held by bearings O and H, as shown in Figure 7.14, rotate at a constant angular velocity

$\omega_1 = 2.0$ rad/s. Suppose the position of the disk shown in the figure is in the XY-plane (Z-axis is perpendicular to this plane) and point B of the strap AB travels as indicated at a constant relative speed $v = 2.0$ m/s. If the angle $\theta = 0$, the length of the strap $AB = 100$ mm, find (a) the velocity at B, and (b) acceleration at B.

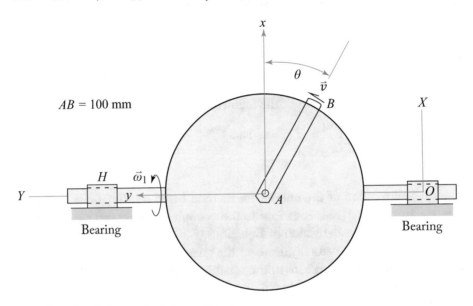

Figure 7.14: Circular disk rotating about Y-axis.

7.2.   A circular plate of uniform thickness and radius $r = 100$ mm rotates at a constant angular speed $\omega_2 = 10$ rad/s with respect to the horizontal arm AO, as shown in Figure 7.15. The arm AO rotates at constant angular speed $\omega_1 = 10$ rad/s about the fixed Z-axis (which coincides with the z-axis of the RFR, as shown in the figure) of the FFR $OXYZ$ (which is known as the Newtonian or inertial frame of reference). For the position shown, determine (a) the velocity and (b) the acceleration of point D with $\vec{r}_{D/A} = 0.10\vec{j}$ m.

7.3.   A crane rotates with a constant angular velocity $\omega_1 = 1$ rad/s as shown in Figure 7.16. The boom is being simultaneously raised with a constant angular $\omega_2 = 0.2$ rad/s relative to the cab to which the origins O of the FFR and RFR are situated. It has been measured that the length of the boom $OP = 20$ m, and the angle $\theta = 40°$. Find (a) the velocity at P and (b) the acceleration at P. Note that the boom may be considered rigid such that it does not bend or extend while in motion.

7.4.   A rectangular plate is welded to a vertical shaft OB which rotates with a constant angular velocity $\vec{\omega}_1$, as shown in Figure 7.17. A rigid beam OA of length r rotates on the smooth surface of the plate and about the origin O with a constant angular velocity $\vec{\omega}_2$ with

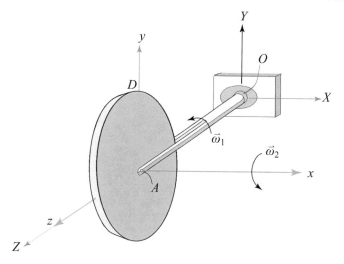

Figure 7.15: Circular plate rotating about Z-axis (dimensions in mm).

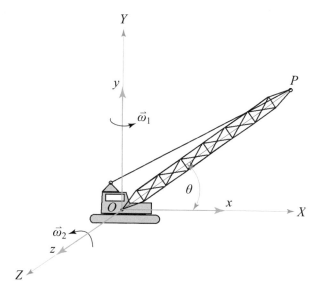

Figure 7.16: A crane in motion.

respect to the rectangular plate. Note that in the figure $OXYZ$ is the *FFR* while $Oxyz$ is the *RFR*. If $r = 200$ mm, $\beta = 45°$, $\theta = 30°$, $\omega_1 = 10$ rad/s, and $\omega_2 = 20$ rad/s, determine

(a) the velocity and

(b) the acceleration at point A.

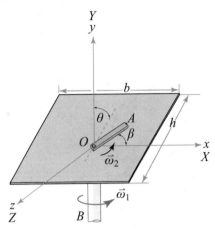

Figure 7.17: Rotating rectangular plate and beam.

7.5. For the linkage shown in Figure 7.18 in which the axes of the *RFR* are indicated with its origin at A and attached to link 3, link 2 rotates clockwise at a constant angular velocity, $\omega_2 = 1.0$ rad/s. The angle between the horizontal line $O_2 O_4$ and link 2 is $\theta_2 = 35.51°$. By applying the unit vectors method, determine the angular velocities $\vec{\omega}_3$, $\vec{\omega}_4$, velocity, and acceleration at point C. Note that the points O_2, A, and C are collinear.

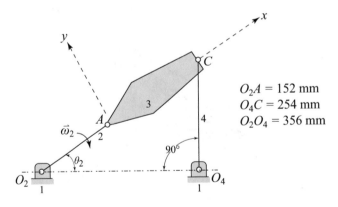

$O_2 A = 152$ mm
$O_4 C = 254$ mm
$O_2 O_4 = 356$ mm

Figure 7.18: A four-bar mechanism.

7.6. For the linkage shown in Figure 7.19, link 2 rotates counter-clockwise at a constant angular velocity, $\omega_2 = 2.0$ rad/s and $\angle AO_2 B = 120°$. By applying the unit vectors method, determine the velocity $\vec{v}_B$ and acceleration $\vec{a}_B$. Note that the axes of the *RFR* with its origin at A and attached to link 3 are indicated in the figure in which dimensions are in mm.

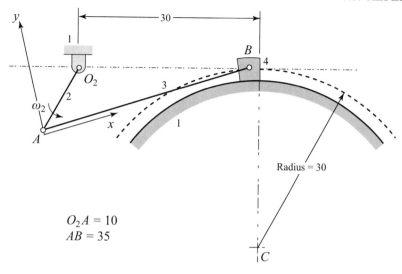

Figure 7.19: Four-bar mechanism with a slider on circular ground (link 1).

REFERENCES

[1] Mabie, H. H. and Ocvirk, F. W., *Mechanisms and Dynamics of Machinery*, 3rd ed.-SI Version, John Wiley & Sons, 1978. 109

[2] Norton, R. L., *Design of Machinery*, 2nd ed., McGraw-Hill, 1999. DOI: 10.1115/1.1605770.

[3] To, C. W. S., *Introduction to Kinematics and Dynamics of Machinery*, Morgan & Claypool Publishers, 2018. DOI: 10.2200/s00798ed1v01y201709mec007. 109, 114, 124

[4] Torby, B. J., *Advanced Dynamics for Engineers*, Holt, Rinehart and Winston, 1984. DOI: 10.1115/1.3269470. 110, 111

[5] Crisfield, M. A., *Nonlinear Finite Element Analysis of Solids and Structures*, Vol. 2, Wiley, 1997. 110

CHAPTER 8

Dynamics of Rigid Bodies

8.1 INTRODUCTION

In the last chapter kinematics of rigid bodies has been presented. In the present chapter the dynamics and kinetics of rigid bodies are considered. Section 8.2 is concerned with the derivation of equations of motion of rigid bodies in 3D space. Equations of motion of rigid bodies in 2D space are presented in Section 8.3. The method of work and energy is introduced in Section 8.4. Impulse, momentum, and angular momentum of rigid bodies are included in Section 8.5. Equations for conservation of momentum and angular momentum are considered in Section 8.6. Section 8.7 has to do with the introduction to impulse motion while eccentric impact of rigid bodies is dealt with in Section 8.8.

8.2 EQUATIONS OF MOTION OF RIGID BODIES IN 3D SPACE

The two most fundamental equations for rigid body dynamics are the force and angular momentum. In this section the equations for translational motion and rotational motion are presented, respectively, in the following two sections. In Section 8.2.3, equations of *constrained motions* are introduced. The *Euler angles* and motion of a *gyroscope* are considered in Section 8.2.4. The final section is concerned with the analysis of *steady precession*.

8.2.1 EQUATIONS OF TRANSLATIONAL MOTION

These equations can be derived similar to those obtained for a system of particles presented in Chapter 6. Specifically, by making use of Equations (6.14) and (6.24), one has the equation for the forces

$$\sum_{i=1}^{n} \vec{F}_i = m\bar{\vec{a}} = m\vec{a}_G, \tag{8.1}$$

where now the term on the lhs is the externally applied forces acting on the rigid body while m is the mass and $\vec{a}_G$, the acceleration at the center of mass G of the rigid body. The symbol $\bar{\vec{a}}$ in Equation (6.24) is replaced by $\vec{a}_G$ to distinguish the accelerations at the mass center of a system of particles there and that in Equation (8.1) of a rigid body, as shown in Figure 8.1 in which $OXYZ$ is the fixed or Newtonian frame of reference (*FFR*).

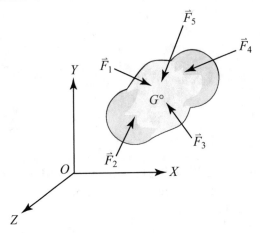

Figure 8.1: Forces acting on a rigid body.

When solving for problems in 3D space it is sometimes more convenient to express the vector equation into three scalar ones as

$$\sum_{i=1}^{n}(F_i)_X = m\,(a_G)_X\,, \quad \sum_{i=1}^{n}(F_i)_Y = m\,(a_G)_Y\,, \quad \sum_{i=1}^{n}(F_i)_Z = m\,(a_G)_Z\,, \qquad (8.2)$$

in which

$$\sum_{i=1}^{n}\vec{F_i} = \sum_{i=1}^{n}(F_i)_X\,\vec{I} + \sum_{i=1}^{n}(F_i)_Y\,\vec{J} + \sum_{i=1}^{n}(F_i)_Z\,\vec{K}$$

is assumed and the subscripts X, Y, and Z denote, respectively, the components along the X-axis, Y-axis, and Z-axis. The symbols $\vec{I}$, $\vec{J}$, $\vec{K}$ designate the unit vectors parallel to the X-axis, Y-axis, and Z-axis, respectively.

Before leaving this section it should be pointed out that $Gx'y'z'$ is applied in the remaining parts of the present chapter as the coordinate system with its origin at the *center of mass* of the rigid body and $\vec{i}$, $\vec{j}$, $\vec{k}$ are the unit vectors along the x', y', z' axes, respectively, while $Oxyz$ is employed to denote the coordinate system in the *RFR*. Thus, the notations for frames of reference employed in this book are summarized in the following, unless it is stated otherwise.

(a) In Chapter 6, $Oxyz$ and associated unit vectors $\vec{i}$, $\vec{j}$, $\vec{k}$ has been employed to denote the *FFR* for systems of particles, while $Gx'y'z'$ is the coordinate system with its origin located at the mass center G of the system of particles.

(b) In Chapter 7, $OXYZ$ and associated unit vectors $\vec{I}$, $\vec{J}$, $\vec{K}$ has been applied to the *FFR*, while $Axyz$ with associated unit vectors $\vec{i}$, $\vec{j}$, $\vec{k}$ is the body attached frame or *RFR*.

(c) In the remaining parts of this chapter, *OXYZ* is applied to denote the *FFR* while *Gxyz* with associated unit vectors $\vec{i}, \vec{j}, \vec{k}$ is used as the coordinate system whose origin is at the mass center *G* of the rigid body.

8.2.2 EQUATIONS OF ROTATIONAL MOTION

Recall that Equation (6.28) for the angular momentum of a system of particles, $\vec{H}_G = \vec{H}'_G$ which indicates that the angular momentum is independent of whether the velocities of the particles are referenced to the origin *O* of *FFR* or to the center of mass *G*. With reference to Equation (6.28) and Figure 8.2, one can write

$$\vec{H}_G = \sum_{i=1}^{n} \vec{r}'_i \times \Delta m_i \vec{v}'_i,$$

where Δm_i is the mass of the particle P_i. It should be mentioned that for a particle no rotation is possible. However, for a rigid body which can be considered as a system with infinite number of particles and in the above equation the summation sign can be replaced by the integral sign and in the limit Δm_i approaches to dm. The velocity $\vec{v}'_i$ can be replaced by the cross product of the angular velocity $\vec{\omega}$ of the body and position vector $\vec{r}'_i$, which is the position vector of the particle P_i relative to the centroidal frame (this is the notation used in Chapter 6). Then the last equation becomes

$$\vec{H}_G = \int \vec{r}'_i \times \left(\vec{\omega} \times \vec{r}'_i\right) dm. \tag{8.3}$$

By making use of the vector triple product $A \times (B \times C) = (A \cdot C) B - (A \cdot B) C$, or by writing $\vec{H}_G = (H_G)_x \vec{i} + (H_G)_y \vec{j} + (H_G)_z \vec{k}$, $\vec{r}'_i = x\vec{i} + y\vec{j} + z\vec{k}$ and $\vec{\omega} = \omega_x \vec{i} + \omega_y \vec{j} + \omega_z \vec{k}$ as well as reference to Equation (3.16), one can write

$$(H_G)_x = \int \left[y\left(\vec{\omega} \times \vec{r}'_i\right)_z - z\left(\vec{\omega} \times \vec{r}'_i\right)_y \right] dm$$

$$= \int \left[y\left(\omega_x y - \omega_y x\right) - z\left(\omega_z x - \omega_x z\right) \right] dm$$

$$= \omega_x \int \left(y^2 + z^2\right) dm - \omega_y \int xy\, dm - \omega_z \int zx\, dm$$

or

$$(H_G)_x = \omega_x \bar{I}_x - \omega_y \bar{I}_{xy} - \omega_z \bar{I}_{xz}, \tag{8.4}$$

where the *centroidal mass moments of inertia* and *centroidal mass products of inertia* of the body are, respectively, $\bar{I}_x = \int \left(y^2 + z^2\right) dm$, $\bar{I}_{xy} = \int xy\, dm$, and $\bar{I}_{xz} = \int zx\, dm$.

Similarly, with reference to Equation (3.16), one can show that

$$(H_G)_y = -\omega_x \bar{I}_{yx} + \omega_y \bar{I}_y - \omega_z \bar{I}_{yz} \tag{8.5}$$

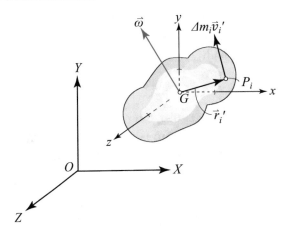

Figure 8.2: Angular momentum relative to centroidal frame.

and

$$(H_G)_z = -\omega_x \bar{I}_{zx} - \omega_y \bar{I}_{zy} + \omega_z \bar{I}_z. \tag{8.6}$$

Equations (8.4)–(8.6) can be written in matrix form as

$$\begin{pmatrix} (H_G)_x \\ (H_G)_y \\ (H_G)_z \end{pmatrix} = \begin{bmatrix} \bar{I}_x & -\bar{I}_{xy} & -\bar{I}_{xz} \\ -\bar{I}_{yx} & \bar{I}_y & -\bar{I}_{yz} \\ -\bar{I}_{zx} & -\bar{I}_{zy} & \bar{I}_z \end{bmatrix} \begin{pmatrix} \omega_x \\ \omega_y \\ \omega_z \end{pmatrix}. \tag{8.7}$$

- It is always possible to choose a system of axes, known as the *principal axes of inertia* of a rigid body such that the products of inertia become zero. That is, the coefficient matrix on the rhs of Equation (8.7) reduces to a diagonal matrix.

- If the three principal centroidal moments of inertia are equal and the corresponding three components of the angular momentum about the mass center G are proportional to the corresponding three components of the angular velocity then the vectors $\vec{H}_G$ and $\vec{\omega}$ are collinear.

Having presented the angular momentum of a rigid body in 3D space, one is ready to consider the equations of rotational motion for a rigid body. The Newtonian or inertia or fixed frame of reference $OXYZ$, and the *body attached frame* or *rotating frame of reference Gxyz*, which is generally chosen so that the moments and products of inertia are independent of time, are shown in Figure 8.3.

With reference to Equation (6.27), the rotational equations of motion for a rigid body is given by

$$\sum_{i=1}^{n} \left(\vec{M}_G \right)_i = \dot{\vec{H}}_G. \tag{8.8}$$

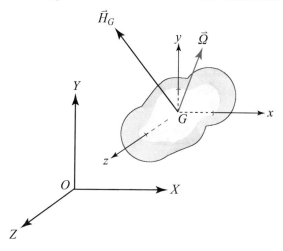

Figure 8.3: A rigid body with body attached frame of reference.

Note in Equation (8.3) $\vec{\omega}$ is the angular velocity of the rigid body. However, in general, the frame of reference $Gxyz$ may be rotating w.r.t. the *FFR* such that the unit vectors $\vec{i}$, $\vec{j}$, $\vec{k}$ are time dependent. Let $\vec{\Omega}$ be the angular velocity of $Gxyz$. Thus, by applying Equation (7.6) in which the vector $\vec{P}$ is replaced by $\vec{H}_G$, one has

$$\left(\dot{\vec{H}}_G\right)_{OXYZ} = \frac{d\vec{H}_G}{dt} = \left(\dot{\vec{H}}_G\right)_{Gxyz} + \vec{\Omega} \times \vec{H}_G, \tag{8.9}$$

where

$\left(\dot{\vec{H}}_G\right)_{OXYZ}$ is the rate of change of $\vec{H}_G$ of the rigid body w.r.t. the *FFR*,

$\left(\dot{\vec{H}}_G\right)_{Gxyz}$ is the rate of change of $\vec{H}_G$ of the rigid body with reference to the *RFR*, and

$\vec{\Omega}$ is the angular velocity of the *RFR*.

It may be appropriate to note that applying Equation (7.6) with the vector $\vec{P}$ replaced by $\vec{\omega}$ one can obtain

$$\left(\dot{\vec{\omega}}\right)_{OXYZ} = \frac{d\vec{\omega}}{dt} = \left(\dot{\vec{\omega}}\right)_{Gxyz} + \vec{\omega} \times \vec{\omega} = \left(\dot{\vec{\omega}}\right)_{Gxyz}.$$

This states the important fact that *the angular acceleration is independent of the frame of reference.* In some problems this result can simplify the steps in the solution.

8.2.3   EQUATIONS OF CONSTRAINED MOTIONS

In the foregoing sections, the equations of motion apply to unconstrained rigid bodies. In this section equations of motion of *constrained rigid bodies* are introduced.

Consider first the case in which the rigid body is constrained to a *fixed point* O. With reference to Figure 8.4 and by making use of Equations (8.8) and (8.9), one can obtain

$$\sum_{i=1}^{n} \left(\vec{M_o} \right)_i = \dot{\vec{H}}_o. \tag{8.10}$$

$$\left(\dot{\vec{H}}_o \right)_{OXYZ} = \left(\frac{d\vec{H}_o}{dt} \right)_{OXYZ} = \left(\dot{\vec{H}}_o \right)_{Gxyz} + \vec{\Omega} \times \vec{H}_o, \tag{8.11}$$

where

$\left(\dot{\vec{H}}_o \right)_{OXYZ}$ is the rate of change of $\vec{H}_o$ of the rigid body w.r.t. the *FFR*,

$\left(\dot{\vec{H}}_o \right)_{Oxyz}$ is the rate of change of $\vec{H}_o$ of the rigid body with reference to the *RFR*, and

$\vec{\Omega}$ is the angular velocity of the *RFR*.

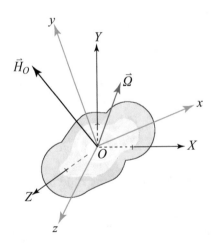

Figure 8.4: A rigid body constrained to a fixed point.

Now, consider the case in which the rigid body is *constrained to rotate about a fixed axis*, as shown in Figure 8.5. For example, one is interested in setting up the equations of motion of a rigid rotor supported by bearings.

With reference to Figure 8.5, one observes that the angular velocity of the body w.r.t. the fixed frame $OXYZ$ is denoted by $\vec{\omega} = \omega \vec{k}$. Substituting $\omega_x = 0, \omega_y = 0$, and $\omega_z = \omega$ into

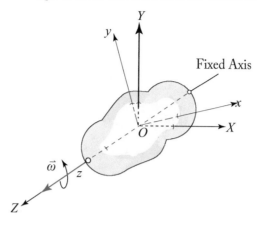

Figure 8.5: A rigid body constrained to a fixed axis.

Equation (8.7) for the present problem, one obtains

$$(H_o)_x = -I_{xz}\omega, \quad (H_o)_y = -I_{yz}\omega, \quad (H_o)_z = I_z\omega.$$

Since $Oxyz$ is attached to the rigid body such that $\vec{\Omega} = \vec{\omega}$, Equation (8.11) gives

$$\left(\dot{\vec{H}}_o\right)_{OXYZ} = \left(\dot{\vec{H}}_o\right)_{Oxyz} + \vec{\omega} \times \vec{H}_o$$

$$= \left(-I_{xz}\vec{\imath} - I_{yz}\vec{\jmath} + I_z\vec{k}\right)\frac{d\omega}{dt} + \omega\vec{k} \times \left(-I_{xz}\vec{\imath} - I_{yz}\vec{\jmath} + I_z\vec{k}\right)\omega$$

$$= \left(-I_{xz}\vec{\imath} - I_{yz}\vec{\jmath} + I_z\vec{k}\right)\alpha + \left(I_{yz}\vec{\imath} - I_{xz}\vec{\jmath}\right)\omega^2$$

$$= \left(-I_{xz}\alpha + I_{yz}\omega^2\right)\vec{\imath} - \left(I_{xz}\omega^2 + I_{yz}\alpha\right)\vec{\jmath} + I_z\alpha\vec{k}.$$

It may be appropriate to note that if ω is constant such that $\frac{d\omega}{dt} = \alpha = 0$ then the term

$$\left(\dot{\vec{H}}_o\right)_{Oxyz} = \left(-I_{xz}\vec{\imath} - I_{yz}\vec{\jmath} + I_z\vec{k}\right)\frac{d\omega}{dt} = 0.$$

However, in general, $\frac{d\omega}{dt} \neq 0$. Therefore, by making use of Equation (8.10) and writing component by component, one has

$$\sum_{i=1}^{n} (M_o)_i^x = -I_{xz}\alpha + I_{yz}\omega^2, \tag{8.12a}$$

$$\sum_{i=1}^{n} (M_o)_i^y = -I_{xz}\omega^2 - I_{yz}\alpha, \tag{8.12b}$$

$$\sum_{i=1}^{n} (M_o)_i^z = I_z\alpha, \tag{8.12c}$$

where the superscripts x, y, and z designate the components along the unit vector directions $\vec{i}$, $\vec{j}$, and $\vec{k}$, respectively, of the *RFR*.

   If the externally applied forces are known in the body the angular acceleration α can be determined by Equation (8.12c). This acceleration α can be integrated to obtain the angular velocity ω. The angular velocity and acceleration can be substituted into Equations (8.12a) and (8.12c). These equations, in addition to those in Equation (8.2), can be applied to find the reactions at the bearings, B_1 and B_2 in Figure 8.5 assuming the body is a rigid rotor supported by the bearings.

Example 8.1
A high-frequency radar used in a frigate can be considered as a rectangular plate of uniform thickness of mass m and side lengths b and h. The plate is welded to a vertical rod AB, as shown in Figure 8.6a. The angle between the vertical rod and the plate is $\theta = 30°$. If the vertical rod rotates at a constant angular velocity $\vec{\omega}$, determine (a) the force and couple system representing the dynamic reaction at point A and (b) the dynamic reaction at A when $\omega = 1$ rad/s anticlockwise, $m = 1,000$ kg, $b = 1.6$ m, and $h = 0.8$ m.

Solution:
This solution requires use of Equations (8.1) and (8.11), $\sum_{i=1}^{n} \vec{F}_i = m\vec{a} = m\vec{a}_G$, and $\left(\dot{\vec{H}}_o\right)_{OXYZ} = \left(\dot{\vec{H}}_o\right)_{Gxyz} + \vec{\Omega} \times \vec{H}_o$, in which the subscript o is replaced by A in the present problem.

Angular momentum $\vec{H}_A$
Consider an auxiliary principal axes $Axy'z'$, as shown in Figure 8.6b such that

$$\omega_x = 0, \quad \omega_{y'} = \omega\cos 30°, \quad \omega_{z'} = \omega\sin 30°,$$

$$I_{y'} = \frac{1}{12}mb^2, \quad I_{z'} = \frac{1}{12}m(b^2 + h^2) + m\left(\frac{h}{2}\right)^2 = \frac{1}{12}mb^2 + \frac{1}{3}mh^2,$$

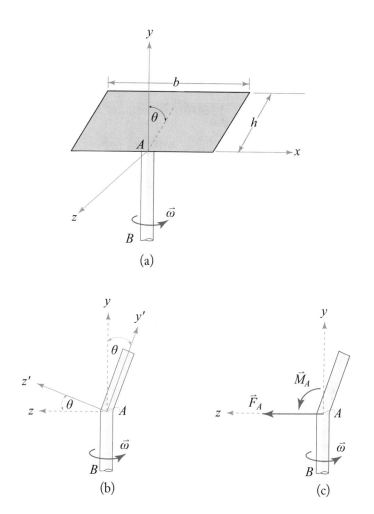

Figure 8.6: (a) Rotating high-frequency rectangular radar in a frigate, (b) auxiliary principal axes, and (c) force and couple system at A.

angular momentum,

$$\vec{H}_A = I_x \omega_x \vec{i} + I_{y'} \omega_{y'} \vec{j}' + I_{z'} \omega_{z'} \vec{k}'$$

$$= 0 + \left(\frac{1}{12} mb^2 \omega \cos 30°\right) \vec{j}' + \left(\frac{1}{12} mb^2 + \frac{1}{3} mh^2\right) \omega \sin 30° \, \vec{k}'.$$

But $\vec{j}' = \cos 30° \, \vec{j} - \sin 30° \, \vec{k}$, and $\vec{k}' = \sin 30° \, \vec{j} + \cos 30° \, \vec{k}$. Therefore,

$$\vec{H}_A = I_x \omega_x \vec{i} + I_{y'} \omega_{y'} \vec{j}' + I_{z'} \omega_{z'} \vec{k}'$$

$$= \left(\frac{1}{12} mb^2 \omega \cos 30°\right) \left(\cos 30° \vec{j} - \sin 30° \vec{k}\right)$$

$$+ \left(\frac{1}{12} mb^2 + \frac{1}{3} mh^2\right) \omega \sin 30° \left(\sin 30° \vec{j} + \cos 30° \vec{k}\right).$$

Simplifying, one has

$$\vec{H}_A = I_x \omega_x \vec{i} + I_{y'} \omega_{y'} \vec{j}' + I_{z'} \omega_{z'} \vec{k}'$$

$$= m\omega \left(\frac{1}{12} b^2 + \frac{1}{3} h^2 \sin^2 30°\right) \vec{j} + m\omega \left(\frac{1}{3} h^2\right) \sin 30° \cos 30° \, \vec{k}. \qquad (8.13)$$

Force and couple system

Since $\vec{\omega}$ is constant such that $\dot{\vec{\omega}} = \frac{d\vec{\omega}}{dt} = \vec{\alpha} = 0$ which leads to $\left(\dot{\vec{H}}_A\right)_{Axyz} = 0$. In addition, the angular velocity of the *RFR*, $\vec{\Omega} = \vec{\omega} = \omega \vec{j}$. Hence, upon application of Equation (8.11),

$$\sum_{i=1}^{n} \left(\vec{M}_A\right)_i = \left(\dot{\vec{H}}_A\right)_{AXYZ} = \left(\dot{\vec{H}}_A\right)_{Axyz} + \vec{\Omega} \times \vec{H}_A = 0 + (\omega \vec{j}) \times \vec{H}_A.$$

Substituting for Equation (8.13) and simplifying, it leads to

$$\sum_{i=1}^{n} \left(\vec{M}_A\right)_i = \vec{M}_A = m\omega^2 \left(\frac{1}{3} h^2\right) \sin 30° \cos 30° \vec{i}. \qquad (8.14a)$$

Application of Equation (8.1), gives

$$\sum_{i=1}^{n} \vec{F}_i = \vec{F}_A = m\vec{a}_A = m\omega^2 \left(\frac{h}{2}\right) \sin 30° \, \vec{k}. \qquad (8.14b)$$

(a) Equations (8.14a) and (8.14b) give the required force and couple system representing the dynamic reaction at point A.

(b) Substituting for the given data, the dynamic force and couple system at A are obtained as

$$\vec{F}_A = m\omega^2 \left(\frac{h}{2}\right) \sin 30° \vec{k} = 200 \vec{k} \text{ N},$$

$$\vec{M}_A = m\omega^2 \left(\frac{1}{3}h^2\right) \sin 30° \cos 30° \vec{\imath} = 92.3760 \vec{\imath} \text{ Nm}.$$

8.2.4   EULERIAN ANGLES AND MOTION OF A GYROSCOPE

A *gyroscope* is generally defined as a rotating body having one axis of symmetry and the rotation about the axis of symmetry is relatively large compared with the rotation about any other axis. The term gyroscope was introduced by Foucault to denote a device that can be applied to prove the movement of the earth [1]. A modern example is a rotor mounted in a Cardan's suspension, as shown in Figure 8.7a. A gyroscope can assume any orientation. However, its mass center must be stationary in space. The two chosen frames of reference are the inertial frame of reference $OXYZ$ and the body fixed or body attached frame $Oxyz$ which is frequently called the *RFR*.

To completely characterize the position of the gyroscope at any given moment, the so-called *Eulerian* or *Euler angles* are used. The commonly adopted notation of Eulerian angles (in radian) in sequence is $\phi, \theta,$ and, ψ where

ϕ is the rotation of the outer gimbal about $O_1 O_2$ (from the reference position in Figure 8.7a to the position in Figure 8.7b),

θ is the rotation of the inner gimbal about $A_1 A_2$, and

ψ is the rotation of the rotor about $B_1 B_2$.

It may be appropriate to mention that the sequence of these three angles must be maintained [1, 2] because *finite rotations are not vectors* (of course, for infinitesimal small rotations they can be expressed as vectors).

Recall, the unit vectors $\vec{\imath}, \vec{\jmath},$ and $\vec{k}$, respectively, along the $x, y,$ and z axes of the *RFR* are attached to the rotor. Note that the y-axis along $A_1 A_2$ and the z-axis along $B_1 B_2$ are principal axes of inertia for the gyroscope. One can write the angular velocity $\vec{\omega}$ of the gyroscope w.r.t. the *FFR* as

$$\vec{\omega} = \dot{\phi}\vec{K} + \dot{\theta}\vec{\jmath} + \dot{\psi}\vec{k} \tag{8.15}$$

in which $\vec{K}$ is the unit vector along the Z-axis of the *FFR* while $\dot{\phi} = \frac{d\phi}{dt}, \dot{\theta} = \frac{d\theta}{dt},$ and $\dot{\psi} = \frac{d\psi}{dt}$ are, respectively, the *precession, nutation,* and *spin*. As the vector components for $\vec{\omega}$ in Equation (8.15) are not orthogonal $\vec{K}$ is expressed in components parallel to the x and z axes. That is,

$$\vec{K} = -\sin\theta \vec{\imath} + \cos\theta \vec{k}.$$

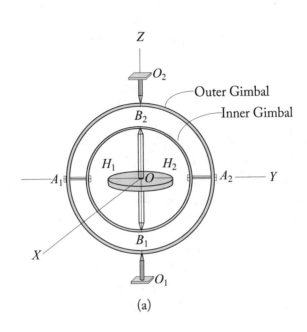

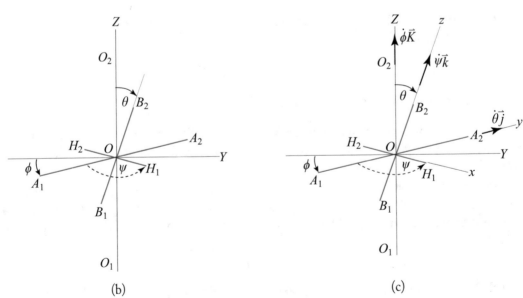

Figure 8.7: Gyroscope with (a) *FFR*, (b) Eulerian angles, and (c) angular velocities.

Substituting this equation into (8.15) and rearranging, one obtains

$$\vec{\omega} = -\dot{\phi}\sin\theta\,\vec{\imath} + \dot{\theta}\,\vec{\jmath} + \left(\dot{\psi} + \dot{\phi}\cos\theta\right)\vec{k} \qquad (8.16)$$

which can be identified with the angular velocity in Equation (8.3),

$$\vec{\omega} = \omega_x\,\vec{\imath} + \omega_y\,\vec{\jmath} + \omega_z\,\vec{k}$$

so that

$$\omega_x = -\dot{\phi}\sin\theta, \quad \omega_y = \dot{\theta}, \quad \omega_z = \dot{\psi} + \dot{\phi}\cos\theta. \qquad (8.17)$$

Thus, if one denotes the moment of inertia of the rotor about its spin axis by I_z, moment of inertia about a transverse axis through mass center O by I_x, and disregarding the mass of the gimbals as well as making use of Equation (8.7), the angular momentum of the gyroscope about O becomes

$$\begin{pmatrix} (H_o)_x \\ (H_o)_y \\ (H_o)_z \end{pmatrix} = \begin{bmatrix} I_x & 0 & 0 \\ 0 & I_x & 0 \\ 0 & 0 & I_z \end{bmatrix} \begin{pmatrix} \omega_x \\ \omega_y \\ \omega_z \end{pmatrix},$$

or in vector form

$$\vec{H}_o = -\left(I_x\dot{\phi}\sin\theta\right)\vec{\imath} + \left(I_x\dot{\theta}\right)\vec{\jmath} + I_z\left(\dot{\psi} + \dot{\phi}\cos\theta\right)\vec{k}. \qquad (8.18)$$

Before the equations of motion for the above gyroscope are derived, one needs to obtain the angular velocity $\vec{\Omega}$ of the *RFR* which, in the present case, is attached to the inner gimbal and therefore,

$$\vec{\Omega} = \dot{\phi}\vec{K} + \dot{\theta}\vec{\jmath}$$

which can be identified as $\vec{\omega}$ with its spin component equal to zero. Thus, by Equation (8.16) one has

$$\vec{\Omega} = -\left(\dot{\phi}\sin\theta\right)\vec{\imath} + \dot{\theta}\vec{\jmath} + \left(\dot{\phi}\cos\theta\right)\vec{k}. \qquad (8.19)$$

Substituting Equations (8.18) and (8.19) into (8.11), and operating on the resulting equation as well as rearranging, one can obtain the equations for the motion of the gyroscope as

$$\sum_{i=1}^{n} (M_o)_i^x = -I_x\left(\ddot{\phi}\sin\theta + 2\dot{\theta}\dot{\phi}\cos\theta\right) + I_z\dot{\theta}\left(\dot{\psi} + \dot{\phi}\cos\theta\right), \qquad (8.20a)$$

$$\sum_{i=1}^{n} (M_o)_i^y = I_x\left(\ddot{\theta} - \dot{\phi}^2\sin\theta\cos\theta\right) + I_z\dot{\phi}\sin\theta\left(\dot{\psi} + \dot{\phi}\cos\theta\right), \qquad (8.20b)$$

$$\sum_{i=1}^{n} (M_o)_i^z = I_z\frac{d\left(\dot{\psi} + \dot{\phi}\cos\theta\right)}{dt}. \qquad (8.20c)$$

The above three equations are highly nonlinear and in order to obtain the solution numerical methods are generally required. This is beyond the scope of the present book and will not be pursued further. However, in the next section a special case known as *steady precession of a gyroscope* will be considered.

- Before leaving this section, it should be noted that Equations (8.20a,b,c) can be applied to describe the motion of an axi-symmetrical body or body of revolution attached at a point on its axis of symmetry.

8.2.5 STEADY PRECESSION OF A GYROSCOPE

As noted above, Equations (8.20a,b,c) are highly nonlinear and is therefore difficult, if not impossible, to solve. However, a special case exists such that simplification and therefore solution is possible. This special case is known as *steady precession*. It occurs when the nutation angle θ, precession $\dot{\phi}$, and spin $\dot{\psi}$ are simultaneously constant. These latter conditions lead Equations (8.20a,b,c) to become

$$\sum_{i=1}^{n} (M_o)_i^x = 0, \qquad \sum_{i=1}^{n} (M_o)_i^z = 0, \tag{8.21a,c}$$

$$\sum_{i=1}^{n} (M_o)_i^y = -I_x \dot{\phi}^2 \sin\theta \cos\theta + I_z \dot{\phi} \sin\theta \left(\dot{\psi} + \dot{\phi} \cos\theta\right), \tag{8.21b}$$

By making use of the definition of ω_z in Equation (8.17), Equation (8.21b) becomes

$$\sum_{i=1}^{n} (M_o)_i^y = -I_x \dot{\phi}^2 \sin\theta \cos\theta + I_z \dot{\phi} \omega_z \sin\theta. \tag{8.22}$$

It is interesting to note that when $\theta = 90°$ which means when the precession axis and spin axis are at right angle to each other, Equation (8.22) reduces to

$$\sum_{i=1}^{n} (M_o)_i^y = I_z \dot{\phi} \dot{\psi}. \tag{8.23}$$

This equation has many applications. For example, it is used in gyroscopes for the stabilization of torpedoes and ships, and in bearing force analysis. The effect of gyroscopic force in the bearings is studied in Example 8.2 [3].

Example 8.2
The rotor of a jet airplane engine is supported by two bearings of distance $AB = 2.1334$ m, as shown in Figure 8.8a. The rotor assembly including compressor, turbine, and shaft has a mass of 1,000 kg and a radius of gyration of 250 mm. Determine the maximum bearing force as the

airplane undergoes a pullout on a circular curve whose radius is 2,000 m at a constant airplane speed of 1,000 km/h and the engine rotor speed of 10,000 rpm. In the solution, include the gyroscopic effect and the effect of centrifugal force due to the pullout.

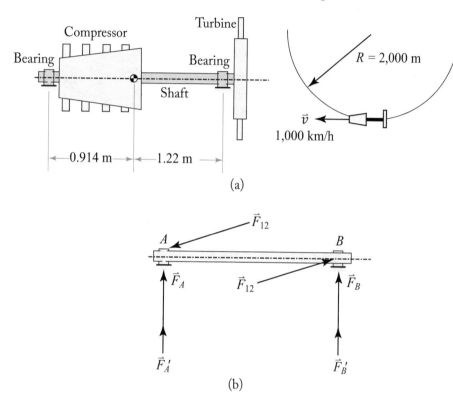

Figure 8.8: Rotor assembly and motion: (a) rotor assembly and motion path and (b) forces at bearings.

Solution:

In obtaining the solution for the present problem, Equation (8.23) will be applied.

Given data

mass of rotor assembly $m = 1{,}000$ kg,

distance between bearings $AB = 2.134$ m,

radius of gyration $\gamma = 250$ mm,

radius of path $R = 2{,}000$ m,

velocity of airplane $v = \dfrac{\left(1{,}000\,\frac{km}{h}\right)\left(1{,}000\,\frac{m}{km}\right)}{\left(3{,}600\,\frac{s}{h}\right)} = 277.78$ m/s, and

rotor rotating speed $\omega = \dfrac{2\pi(10{,}000)}{60}\,\frac{rad}{s} = 1047.20\,\frac{rad}{s}$.

The direction of rotation is ccw, viewed from the rhs in Figure 8.8a.

Gyroscopic force

Apllying Equation (8.23), $\sum_{i=1}^{n} (M_o)_i^y = I_z \dot{\phi} \dot{\psi}$ where $I_z = m\gamma^2$ is the moment of inertia of the assembly, $\dot{\phi} = \omega_p$ the precession, and $\dot{\psi} = \omega_s$ the spin. Thus, the moment about the mass center

$$M_G = I_z \omega_p \omega_s$$

where $I_z = m\gamma^2 = 1{,}000 \, (250/1{,}000)^2 \text{ kg m}^2 = 62.5 \text{ kg m}^2,$

$$\omega_p = \frac{v}{R} = \frac{277.78}{2{,}000} \frac{\text{rad}}{\text{s}} = 0.1389 \frac{\text{rad}}{\text{s}}, \quad \omega_s = \omega = 1047.20 \frac{\text{rad}}{\text{s}}.$$

Therefore, $M_G = I_z \omega_p \omega_s = 62.5 \, (0.1389) \, (1047.20) \text{ Nm} = 9090.35 \text{ Nm}.$ The bearing force on the shaft in the horizontal plane is given by

$$F_{12} = M_G/(AB) = 9090.35/2.134 \text{ N} = 4259.77 \text{ N}.$$

Centrifugal force

The centrifugal force $F = m \left(\frac{v^2}{R} \right) = 1{,}000 \left(\frac{277.78^2}{2{,}000} \right) \text{ N} = 38{,}580.86 \text{ N}.$ Force on bearing A on shaft upward $F_A = \frac{1.22(38{,}580.86)}{AB} \text{ N} = 22056.54 \text{ N}.$ Since the force F is acting downward in the vertical plane (force on rotor and shaft on bearings), $F_B = F - F_A$, upward in the vertical plane.

Therefore $F_B = 16524.32 \text{ N}$, upward in the vertical plane.

Bearing forces due to weight of rotor assembly

$$F_A' = \frac{1.22 \, (1{,}000) \, (9.81)}{AB} \text{ N} = \frac{1.22 \, (1{,}000) \, (9.81)}{2.134} \text{ N} = 5608.34 \text{ N},$$

upward in the vertical plane (force of bearing A on shaft). Bearing force F_B' is given by

$$F_B' = mg - F_A' = 1{,}000 \, (9.81) - 5608.34 \text{ N} = 4201.66 \text{ N},$$

upward (force of bearing B on shaft).
Therefore, $F_A + F_A' = 22056.54 + 5608.34 \text{ N} = 27{,}664.88 \text{ N}$, upward.
Similarly, $F_B + F_B' = 16524.32 + 4201.66 \text{ N} = 20{,}725.98 \text{ N}$, upward.

Total bearing forces

Total bearing force at A,

$$F_{TA} = \sqrt{\left(F_A + F_A' \right)^2 + F_{12}^2} \text{ N}$$

$$= \sqrt{27{,}664.88^2 + (4259.77)^2} \text{ N}$$

$$= 27{,}990.91 \text{ N}.$$

Similarly, total bearing force at B,

$$
\begin{aligned}
F_{TB} &= \sqrt{\left(F_B + F_B'\right)^2 + F_{12}^2} \text{ N} \\
&= \sqrt{20{,}72598^2 + (4259.77)^2} \text{ N} \\
&= 21{,}159.20 \text{ N}.
\end{aligned}
$$

The forces and directions are illustrated in Figure 8.8b.

8.3 EQUATIONS OF MOTION OF RIGID BODIES IN 2D SPACE

Having obtained the equations of motion for rigid bodies in 3D space, it is relatively straight forward to simplify them for rigid body motion in 2D space or commonly called *planar motion* of rigid bodies.

Consider the 2D rigid body, shown in Figure 8.9 and by making use of Equations (8.2) as well as (8.12c), one obtains

$$
\sum_{i=1}^{n} (F_i)_x = m (a_G)_x , \quad \sum_{i=1}^{n} (F_i)_y = m (a_G)_y , \quad \sum_{i=1}^{n} (M_G)_i^z = \bar{I} \alpha, \tag{8.24}
$$

where $\bar{I}$ is the moment of inertia of the rigid body about an axis perpendicular to the plane of the 2D rigid body and through the mass center G. Equations in (8.24) state that *the motion of the 2D rigid body is completely defined by the resultant and moment resultant about mass center G of the external forces exerting on it.*

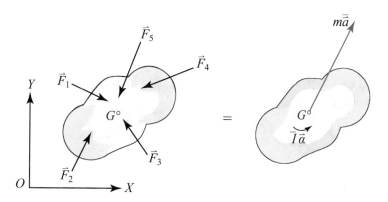

Figure 8.9: A rigid body in 2D space.

If the body is constrained at O to rotate about a fixed axis perpendicular to the plane of the body, as shown in Figure 8.10, the equations of motion become

$$\sum_{i=1}^{n} (F_i)_n = m \, (a_G)_n = m\omega^2 r_G, \tag{8.25a}$$

$$\sum_{i=1}^{n} (F_i)_t = m \, (a_G)_t = m\alpha r_G, \tag{8.25b}$$

$$\sum_{i=1}^{n} (M_G)_i^z = I_G \alpha, \tag{8.25c}$$

in which $(a_G)_n$ is the normal component and $(a_G)_t$ the tangential component of the acceleration $\vec{a}_G$ or $\bar{\vec{a}}$ while I_G or $\bar{I}$ is the mass moment of inertia of the 2D body about the fixed axis of rotation through G.

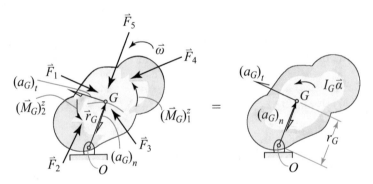

Figure 8.10: Rotation of a 2D body about a fixed axis.

8.4 WORK AND ENERGY IN RIGID BODIES

For rigid bodies, the work can be derived by making use of Equation (4.1a) in which $d\vec{r}$ for a particle is replaced by $d\vec{\theta} \times \vec{r}_i$ for a point P_i of the rigid body. Thus, the incremental work for a rigid body becomes

$$dU = \sum_{i=1}^{n} \vec{F}_i \cdot \left(d\vec{\theta} \times \vec{r}_i \right),$$

where $\vec{r}_i$ is the position vector between the mass center and point P_i.

By integration,

$$U_{1 \to 2} = \int_{\theta_1}^{\theta_2} \sum_{i=1}^{n} \vec{F}_i \cdot \left(d\vec{\theta} \times \vec{r}_i \right)$$

which upon making use of vector algebra reduces to

$$U_{1\to 2} = \int_{\theta_1}^{\theta_2} \sum_{i=1}^{n} \vec{F}_i \cdot \left(d\vec{\theta} \times \vec{r}_i \right) = \int_{\theta_1}^{\theta_2} \sum_{i=1}^{n} \left(\vec{r}_i \times \vec{F}_i \right) \cdot d\vec{\theta}$$

where, by definition, $\vec{r}_i \times \vec{F}_i = \vec{M}_i$ is the moment of external force $\vec{F}_i$ about the center of mass. Therefore,

$$U_{1\to 2} = \int_{\theta_1}^{\theta_2} \sum_{i=1}^{n} \vec{M}_i \cdot d\vec{\theta} = \sum_{i=1}^{n} M_i \left(\theta_2 - \theta_1 \right) \tag{8.26}$$

in which $U_{1\to 2}$ is the *rotational work*.

Remarks:

- The form of the work energy principle for rigid bodies is similar to that in Section 4.3 for a particle. The only thing to remember is that, in general, for rigid bodies the work and energy involve with translational and rotational motions.

- In planar motion, the kinetic energy for a rigid body can be expressed as

$$T_k = \frac{1}{2}m\bar{v}_k^2 + \frac{1}{2}\bar{I}\omega_k^2, \qquad k = 1, 2, \tag{8.27}$$

  where $\bar{v}$ and $\bar{I}$ denote, respectively the velocity at the mass center and mass moment of inertia of the body about the axis through mass center G, while ω is the angular velocity of the body. The subscript k denotes the stage of the motion.

- For 3D motion the rotational component of the kinetic energy is more tedious. In particular, the rotational component of the kinetic energy of the body relative to centroidal axes can be shown to be

$$\begin{aligned} T_r = \frac{1}{2} \big(& \bar{I}_x \omega_x^2 + \bar{I}_y \omega_y^2 + \bar{I}_z \omega_z^2 \\ & -2\bar{I}_{xy}\omega_x\omega_y - 2\bar{I}_{yz}\omega_y\omega_z - 2\bar{I}_{zx}\omega_z\omega_x \big). \end{aligned} \tag{8.28}$$

8.5 IMPULSE, MOMENTUM, AND ANGULAR MOMENTUM OF RIGID BODIES

Similar to that in Section 5.3, an *impulse* is defined as one acting on a rigid body during a very short period of time which is significant enough to produce a define change in momentum. In rigid bodies aside from the *linear impulse* there is the *angular impulse*. For simplicity, in this section impulse, momentum, and angular momentum are confined to 2D space.

With reference to Section 5.2, the equation of translational motion for a rigid body of constant mass m can be written as

$$\sum_{i=1}^{n} \vec{F}_i = m\vec{a}_G = \frac{d(m\vec{v}_G)}{dt},$$

where $m\vec{v}_G$ is the linear momentum and $\vec{v}_G$ is the velocity at the mass center G. One can multiply both sides of this equation by dt and integrating between times t_1 and t_2 such that

$$\int_{t_1}^{t_2} \sum_{i=1}^{n} \vec{F}_i \, dt = \int_{(\vec{v}_G)_1}^{(\vec{v}_G)_2} d(m\vec{v}_G) = m\left(\vec{v}_G\right)_2 - m\left(\vec{v}_G\right)_1$$

or upon re-arranging, one has

$$m\left(\vec{v}_G\right)_1 + \int_{t_1}^{t_2} \sum_{i=1}^{n} \vec{F}_i \, dt = m\left(\vec{v}_G\right)_2. \tag{8.29}$$

This equation is called the *principle of linear impulse and momentum* for a rigid body. It simply states that *the final momentum of a rigid body acting on by a system of impulses is equal to the vector sum of its initial momentum and the impulses during the time interval in question.*

Similarly, the equation of rotational motion from Equation (8.25c) is

$$\sum_{i=1}^{n} (M_G)_i^z = I_G \alpha = I_G \frac{d\omega}{dt}.$$

Multiplying both sides of this equation by dt and integrating between times t_1 and t_2 such that

$$\int_{t_1}^{t_2} \sum_{i=1}^{n} (M_G)_i^z \, dt = \int_{\omega_1}^{\omega_2} I_G \, d\omega = I_G \omega_2 - I_G \omega_1$$

or upon re-arranging, one has

$$I_G \omega_1 + \int_{t_1}^{t_2} \sum_{i=1}^{n} (M_G)_i^z \, dt = I_G \omega_2, \tag{8.30}$$

where the second term on the lhs is interpreted as the system of angular impulses. Equation (8.30) is called the *principle of angular impulse and momentum* for a rigid body. It simply states that *the final angular momentum of a rigid body acting on by a system of angular impulses is equal to the vector sum of its initial angular momentum and the system of impulses during the time interval in question.*

Equations (8.29) and (8.30) are illustrated in Figure 8.11.

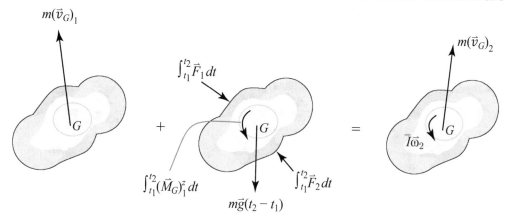

Figure 8.11: Illustration of principles of impulses and momentum.

8.6 CONSERVATION OF MOMENTUM AND ANGULAR MOMENTUM

Returning to Equations (8.29), one observes that if there is no impulse of the external force acting on the body it becomes

$$m\left(\vec{v}_G\right)_1 = m\left(\vec{v}_G\right)_2. \tag{8.31}$$

This equation states that *the total linear momentum of the system is conserved in every direction.* Similarly, if there is no angular impulse acting on the body Equation (8.30) reduces to

$$I_G\omega_1 = I_G\omega_2. \tag{8.32}$$

This equation states that *the angular momentum of the system is conserved.*

It should be noted that in many problems the angular momentum is conserved while the linear momentum is not conserved. Problems in which angular momentum is conserved can be dealt with by the general method of impulse and momentum.

Example 8.3

A uniform rigid beam AB of mass m is suspended from two identical extensible cables, C_1 and C_2 as shown in Figure 8.12a. If for some unknown reasons cable C_2 breaks, find

(a) the angular acceleration $\vec{\alpha}$ of the rigid beam at the instant when C_2 breaks,

(b) the acceleration at point A, and

(c) the acceleration at point B.

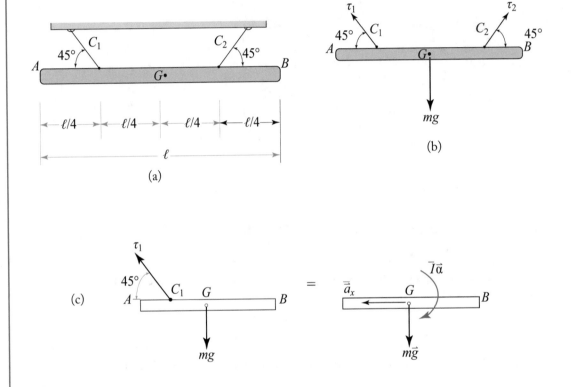

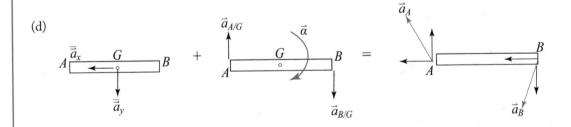

Figure 8.12: Rigid beam suspended from two extensible cables: (a) beam, (b) tensions in cables, (c) FBD when tension is unchanged, and (d) beam in translation and rotation motions.

Solution:

Before any of the cable breaks the tensions τ_1 and τ_2 in the cables are as shown in Figure 8.12b. Thus, summing forces along the vertical direction, one has

$$+\uparrow \sum \vec{F}_y = 0: \qquad \tau_1 \sin 45° + \tau_2 \sin 45° - mg = 0.$$

By symmetry, $\tau_1 = \tau_2$, therefore

$$2\tau_1 \sin 45° - mg = 0$$

which gives

$$\tau_1 = \frac{mg}{2\sin 45°}, \qquad \vec{\tau}_1 = \frac{mg}{2\sin 45°} \qquad \nwarrow 45°.$$

Immediately after cable C_2 breaks the extension or elongation of cable is not altered. Therefore, the tension in cable C_1 remains unchanged. That is, as shown in Figure 8.12c,

$$\vec{\tau}_1 = \frac{mg}{2\sin 45°} \qquad \nwarrow 45°.$$

(a) *Angular acceleration*

Taking moment about mass center G, $\quad +\circlearrowleft \sum \vec{M}_G = \sum \left(\vec{M}_G\right)_e$: Writing $\ell = L$,

$$-\left(\frac{mg \sin 45°}{2 \sin 45°}\right)\frac{L}{4} = \bar{I}\alpha = \left(\frac{mL^2}{12}\right)\alpha \quad \Rightarrow \quad \alpha = -\left(\frac{3}{2}\right)\frac{g}{L} \quad \Rightarrow \quad \vec{\alpha} = \left(\frac{3}{2}\right)\frac{g}{L}\,\circlearrowright.$$

(b) *Acceleration at A*

With reference to Figure 8.12c, and resolving forces along x-axis,

$$+\rightarrow \sum \vec{F}_x = \sum \left(\vec{F}_x\right)_e :$$

$$\frac{-mg}{2\sin 45°}\cos 45° = m\bar{a}_x \quad \Rightarrow \quad \bar{a}_x = -\frac{g}{2} \quad \Rightarrow \quad \vec{\bar{a}}_x = \frac{g}{2} \quad \leftarrow.$$

Resolving forces along y-axis,

$$+\uparrow \sum \vec{F}_y = \sum \left(\vec{F}_y\right)_e :$$

$$-mg + \frac{mg}{2\sin 45°}\sin 45° = m\bar{a}_y \quad \Rightarrow \quad \bar{a}_y = -\frac{g}{2} \quad \Rightarrow \quad \vec{\bar{a}}_y = \frac{g}{2} \quad \downarrow.$$

By making use of the results above and reference to Figure 8.12d, one can write

$$\vec{a}_A = \vec{a}_G + \vec{a}_{A/G} = \vec{\bar{a}}_x + \vec{\bar{a}}_y + \vec{\alpha} \times \vec{r}_{A/G}$$

$$\Rightarrow \quad \vec{a}_A = \left(\frac{g}{2} \leftarrow\right) + \left(\frac{g}{2} \downarrow\right) + \left[\left(\frac{3}{2}\right)\frac{g}{L} \circlearrowright\right] \times \left(\frac{L}{2} \leftarrow\right)$$

$$\Rightarrow \quad \vec{a}_A = \left(\frac{g}{2} \leftarrow\right) + \left(\frac{g}{2} \downarrow\right) + \left[\left(\frac{3}{4}\right)g \uparrow\right]$$

$$\Rightarrow \quad \vec{a}_A = \left(\frac{g}{2} \leftarrow\right) + \left[\left(\frac{1}{4}\right)g \uparrow\right].$$

(c) *Acceleration at B*

The acceleration at B can be similarly obtained as that for $\vec{a}_A$. Thus,

$$\vec{a}_B = \vec{a}_G + \vec{a}_{B/G} = \overline{\vec{a}}_x + \overline{\vec{a}}_y + \vec{\alpha} \times \vec{r}_{B/G}$$

$$\Rightarrow \quad \vec{a}_B = \left(\frac{g}{2} \leftarrow\right) + \left(\frac{g}{2} \downarrow\right) + \left[\left(\frac{3}{2}\right)\frac{g}{L} \circlearrowright\right] \times \left(\frac{L}{2} \rightarrow\right)$$

$$\Rightarrow \quad \vec{a}_B = \left(\frac{g}{2} \leftarrow\right) + \left(\frac{g}{2} \downarrow\right) + \left[\left(\frac{3}{4}\right)g \downarrow\right]$$

$$\Rightarrow \quad \vec{a}_B = \left(\frac{g}{2} \leftarrow\right) + \left[\left(\frac{5}{4}\right)g \downarrow\right].$$

Example 8.4

A homogeneous wheel of mass m and radius r, as shown in Figure 8.13a, is placed on a horizontal surface. It has no linear velocity but has a clockwise angular velocity ω_1 just before it is placed on the surface. It is known that the coefficient of friction between the wheel and the surface is μ. Determine

(a) the time t_2 at which the wheel will begin rolling without sliding and

(b) the linear and angular velocities of the wheel at t_2.

Solution:

Since the wheel is homogeneous the mass moment of inertia about the mass center is

$$\bar{I} = \frac{1}{2}mr^2.$$

The *FBD* is shown in Figure 8.13b. With reference to the *FBD*, the principle of impulse and momentum can be applied.

Principle of impulse and momentum (Figure 8.13b)

Equation (8.30) is employed such that

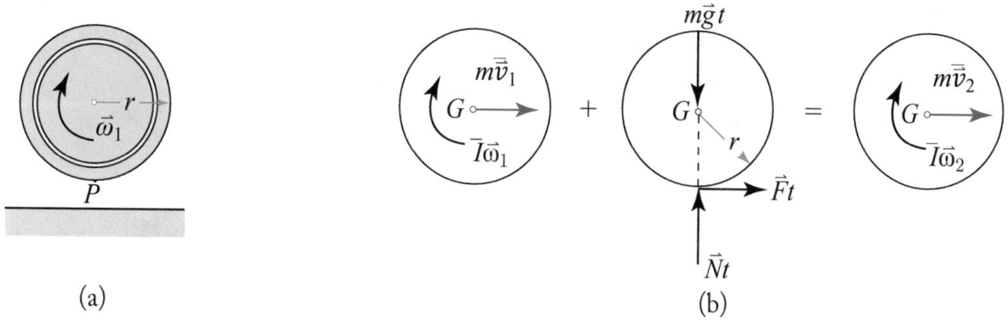

Figure 8.13: (a) Rotating wheel before touching horizontal surface and (b) impulse and momentum.

+ ↑ y components:

$$Nt - mgt = 0. \tag{8.33a}$$

+ → x components:

$$Ft = m\bar{v}_2. \tag{8.33b}$$

+ ↺ moments about G:

$$-\bar{I}\omega_1 + Frt = -\bar{I}\omega_2. \tag{8.33c}$$

From Equation (8.33a), one has

$$N = mg.$$

When $t < t_2$, sliding occurs at point of contact P so that the frictional force

$$F = \mu N = \mu mg.$$

Substituting this into Equation (8.33b), one obtains

$$\bar{v}_2 = \mu gt. \tag{8.33d}$$

Substituting $\bar{I}$ and F into Equation (8.33c), it becomes

$$-\frac{1}{2}mr^2\omega_1 + \mu mgrt = -\frac{1}{2}mr^2\omega_2$$

which gives

$$\omega_2 = \omega_1 - \frac{2\mu gt}{r}. \tag{8.33e}$$

The wheel will begin rolling without sliding when the velocity at the contact point P is zero. That is, P becomes the instantaneous center of rotation at $t = t_2$. Thus, $\bar{v}_2 = r\omega_2$ at $t = t_2$.

By making use of Equations (8.33d) and (8.33e), one obtains

$$\mu g t_2 = r\left(\omega_1 - \frac{2\mu g t_2}{r}\right) \quad \Rightarrow \quad t_2 = \frac{r\omega_1}{3\mu g}.$$

Substituting into Equation (8.33d) and remembering that $t = t_2$ there, it leads to

$$\bar{v}_2 = \mu g r \omega_1 / (3\mu g) = r\omega_1/3.$$

Therefore,

$$\omega_2 = \frac{\bar{v}_2}{r} = \frac{\omega_1}{3}.$$

(a) The time $t_2 = \frac{r\omega_1}{3\mu g}$.

(b) The linear and angular velocities of the wheel at t_2 are, respectively,

$$\vec{\bar{v}}_2 = \frac{r\omega_1}{3} \rightarrow \quad \text{and} \quad \vec{\omega}_2 = \frac{\omega_1}{3} \circlearrowleft .$$

8.7   IMPULSIVE MOTION

For impulsive motion in rigid body, Equations (8.29) and (8.30) can be applied. While these equations are applied to rigid bodies in planar motion they can be generalized for application to rigid bodies in 3D motion. For example, the velocity terms in Equation (8.29) should be understood to be those for every coordinate in the 3D space. Likewise, the angular velocity terms and angular impulse term in Equation (8.30) are referring to those about every coordinate in the 3D space.

To better understand the above comments on impulsive motion the following simple problem is presented.

Example 8.5
A motionless billiard ball has a mass m and radius r resting on a surface, as shown in Figure 8.14. The mass center is at G. If a horizontal impulse $I_{1\to2}$ is applied at a vertical distance ℓ from the surface, determine ℓ so that the billiard ball undergoes pure rolling without slip.

Solution:
Let P be the point of contact between the billiard ball and surface. The principle of linear impulse and momentum, Equation (8.29) and the principle of angular impulse and momentum, Equation (8.30) can be applied.

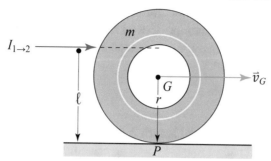

Figure 8.14: Billiard ball with a horizontal impulse.

Principle of linear impulse and momentum

By Equation (8.29) the vertical component is zero since the billiard ball is assumed to remain in contact with the surface. On the other hand, the horizontal component is

$$0 + I_{1\to2} = m\,(v_G)_2\,.$$

This gives

$$(v_G)_2 = \frac{I_{1\to2}}{m}.$$

Principle of angular impulse and momentum

By Equation (8.30) for angular momentum about G,

$$0 + I_{1\to2}\,(\ell - r) = I_G\omega_2 = \frac{2}{5}mr^2\omega_2.$$

Substituting for $I_{1\to2}$, one has

$$m\,(v_G)_2\,(\ell - r) = \frac{2}{5}mr^2\omega_2.$$

Kinematics

The no-slip condition at point P gives $(\vec{v}_G)_2 - r\omega_2 = 0$.
Substituting this value of $(\vec{v}_G)_2$ into the equation of angular impulse and momentum in the foregoing, one obtains

$$\ell = 7r/5.$$

8.8   ECCENTRIC IMPACT

For two rigid bodies in space eccentric impact can occur. The forms of the equations governing the impact motion are similar to those for oblique central impact presented in Chapter 5.

Specifically, for the two rigid bodies undergoing eccentric impact Equations (5.20) can be applied together with one or more equations derived from application of the principle of impulse and momentum. An example illustrating the steps of applying the equations is given in the following.

Example 8.6

A uniform rigid rod of mass 12 kg and length 1.8 m hinged at O is initially at rest, as shown in Figure 8.15a. A spherical ball of mass $m_b = 3$ kg travelling horizontally to the right with an initial velocity of 10 m/s strikes the rod at 1/4 of its length from the free-end A. Knowing that the coefficient of restitution between the rod and sphere is 0.75, determine

(a) the angular velocity ω' of the rod and

(b) the velocity of the ball v'_b immediately after the impact.

Solution:

Let the axes be positive as indicated in Figure 8.15a and mass of the rod m_r as well as velocity of the rod at point B immediately after the impact v'_r. In this problem the rod and the ball can be considered as a single system. When the ball strikes the rod the only external impulsive force applied to the system is the impulsive reaction at O.

Coefficient of restitution
One can write

$$v'_r - v'_b = e\,(v_b - v_r).$$

Since the initial velocity of the ball is given as $v_b = 10$ m/s, the rod is at rest ($v_r = 0$), and $e = 0.75$, therefore

$$v'_r - v'_b = 0.75\,(10 - 0)\ \text{m/s} = 7.5\ \text{m/s}. \tag{8.34a}$$

Kinematics
Since the rod rotates about O, therefore

$$v'_r = (0.75 \times 1.8\ \text{m})\,\omega', \quad \bar{v}'_r = (0.5 \times 1.8\ \text{m})\,\omega'. \tag{8.34b}$$

Principle of angular impulse and momentum (Figure 8.15b)
$+\ \circlearrowleft$ moments about O,

$$I_G\omega + m_b v_b\,(0.75 \times 1.8\ \text{m}) + m_r \bar{v}_r\,(0.5 \times 1.8\ \text{m}) =$$
$$I_G\omega' + m_b v'_b\,(0.75 \times 1.8\ \text{m}) + m_r \bar{v}'_r\,(0.5 \times 1.8\ \text{m}).$$

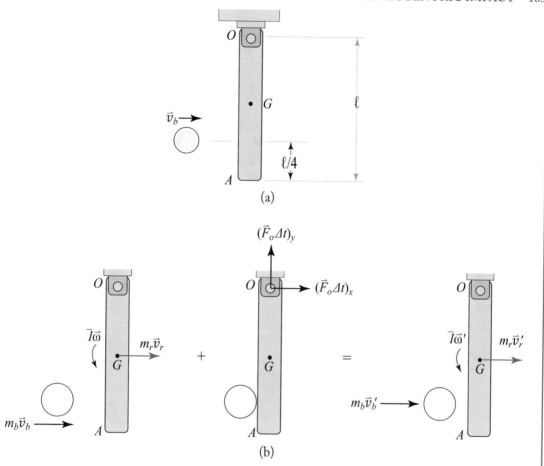

Figure 8.15: (a) A sphere strikes at a hinged rod and (b) impulse and momentum.

Since $\omega = 0$, $\bar{v}_r = 0$, $\bar{v}'_r = 0.5 \times 1.8\omega'$, $I_G = \frac{1}{12}\,(12\text{ kg})\,(1.8\text{ m})^2$, therefore

$$m_b v_b\,(0.75 \times 1.8\text{ m})$$
$$= (1\text{ kg})\,(1.8\text{ m})^2\,\omega' + m_b v'_b\,(0.75 \times 1.8\text{ m}) + m_r \bar{v}'_r\,(0.5 \times 1.8\text{ m})\,.$$

Substituting for $m_b = 3$ kg, $m_r = 12$ kg, $v_b = 10$ m/s, therefore

$$3\,(10)\,(0.75 \times 1.8) = (1.8)^2\,\omega' + 3v'_b\,(0.75 \times 1.8) + 12\omega'\,(0.5 \times 1.8)^2\,.$$

Simplifying,

$$40.5 = 12.96\omega' + 4.05v'_b. \tag{8.34c}$$

Solving Equations (8.34a)–(8.34c), one finds

$$\omega' = 3.8462 \text{ rad/s}, \quad \vec{\omega}' = 3.8462 \text{ rad/s} \quad \circlearrowleft,$$
$$v'_b = -2.3076 \text{ m/s}, \quad \vec{v}'_b = 2.3076 \text{ m/s} \quad \leftarrow.$$

8.9   EXERCISES

8.1. A high-frequency radar used in a frigate can be considered as a square plate of uniform thickness of mass m and side length b. The plate is welded to a vertical rod AB, as shown in Figure 8.16. The angle between the vertical rod and the plate is $\theta = 40°$. If the vertical rod rotates at a constant angular velocity $\vec{\omega}$, determine

(a) the force and couple system representing the dynamic reaction at A,

(b) the dynamic reaction at A when $\omega = 5$ rad/s anti-clockwise, $m = 1{,}000$ kg, $b = 1.6$ m, and

(c) repeat (a) and (b) when now $\theta = 90°$.

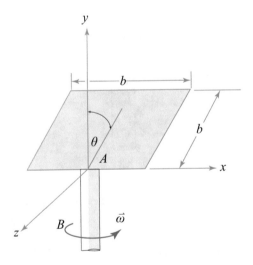

Figure 8.16: Rotating high-frequency square radar.

8.2. A slender uniform rigid rod DA of length $L = 2$ m and mass $m = 15$ kg is pinned at A to a vertical axle BAE which rotates with a uniform angular velocity $\omega = 10$ rad/s. The rod is maintained in the horizontal position, shown in Figure 8.17, by means of a wire DC attached at D of the rod and C of the axle. Given that $\theta = 45°$, find the tension in the wire and the reaction at the pin A.

8.3. A uniform rigid beam AB of mass m is suspended from two identical extensible cables, C_1 and C_2 as shown in Figure 8.18. If for some unknown reasons cable C_2 breaks, find

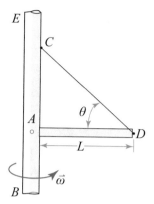

Figure 8.17: Rotating vertical axle with a horizontal rod.

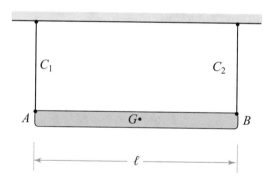

Figure 8.18: Rigid beam suspended from two extensible cables.

(a) the angular acceleration $\vec{\alpha}$ of the rigid beam at this instant,

(b) the acceleration at point A, and

(c) the acceleration at point B.

8.4. In a shooting target practice a uniform square plate has mass m hanging by two strip cables C_1 and C_2 that are attached to the ceiling, as shown in Figure 8.19. The square plate is hit at E by a bullet in a direction perpendicular to the plane of the plate. Suppose the impulse imparted at E is $\vec{F}\Delta t$, find immediately after the impact (a) the velocity of the mass center G and (b) the angular velocity of the plate.

8.5. A uniform circular disk of radius r and mass m is mounted on an axle AG of length ℓ, as shown in Figure 8.20. The axle rotates at a constant angular velocity ω_2 with respect to the vertical shaft AO which rotates at a constant angular velocity ω_1 with respect to the fixed vertical axis. Assume the length of the shaft AO is equal to the radius of the circular

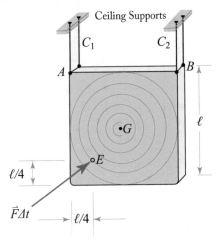

Figure 8.19: Square plate in target practice.

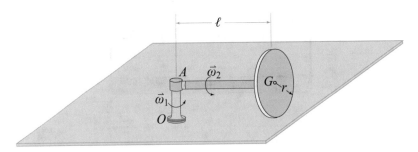

Figure 8.20: Circular disk rolling on a horizontal floor.

disk, and the masses of the axle and shaft can be disregarded, determine (a) the force (for simplicity, it is assumed that this force is vertical) exerted by the floor on the disk (assuming the disk rolls on the horizontal floor without slipping) and (b) the dynamic reaction at point O.

8.6. A uniform thin circular disk of mass m and radius r rolls with a constant spin $\dot{\psi}$ on the horizontal floor, as shown in Figure 8.21. The center of mass G of the disk describes a circular path of radius r_G. If the disk maintains a constant angle of inclination θ during its motion, derive the equation of motion for this particular event. Describe the motion of the disk if the precession $\dot{\phi} = 0$ and $\theta = 90°$ (that is, when the axis of $\dot{\psi}$ in Figure 8.21 is horizontal).

8.7. The angular velocity of a 1,500 kg space capsule is $\vec{\omega}_c = 0.02\vec{\imath} + 0.10\vec{\jmath}$ rad/s when two small jets are activated at A and B, each in a direction parallel to the z-axis. This

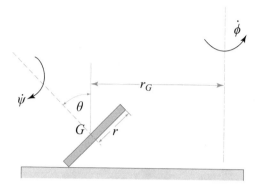

Figure 8.21: Thin disk rolling on a horizontal floor.

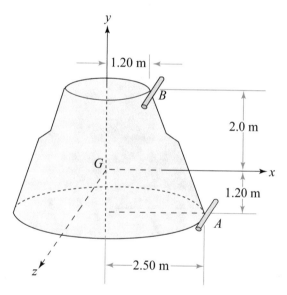

Figure 8.22: Space capsule with two small jets.

capsule is shown in Figure 8.22. Knowing that the radii of gyration of the capsule are $\gamma_x = \gamma_z = 1.00$ m and $\gamma_y = 1.20$ m, and that each of the jets generates a thrust of 50 N, determine (a) the required operation time of each jet if the angular velocity of the capsule is to be reduced to zero and (b) the resulting change in the velocity of the mass center G.

8.8. The conceptual model of a type of aircraft turn indicator is shown in Figure 8.23. Springs AC and BD are initially stretched and exert equal vertical forces at A and B when the path of the airplane is straight. If the rotating disk has a mass of 0.20 kg and spins

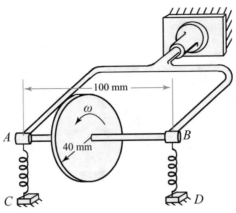

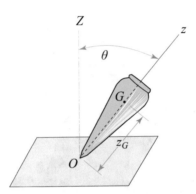

Figure 8.23: Conceptual model of an aircraft turn indicator.

Figure 8.24: A top supported at fixed point O.

at 5,000 rpm, determine the angle through which the yoke rotates when the aircraft executes a horizontal turn of radius 800 m at a speed of 500 km/h. Given that each spring constant is 2,000 N/m.

8.9. A top is supported at the fixed point O, as shown in Figure 8.24. Suppose the moments of inertia of the top about its axis of symmetry and about a transverse axis through O are, respectively, I and I_x, show that the condition for steady precession is

$$\left(I\omega_z - I_x \cos\theta \frac{d\phi}{dt}\right)\frac{d\phi}{dt} = mgz_G,$$

where $\frac{d\phi}{dt}$ is the precession and ω_z is the component of the angular velocity along the axis of symmetry of the top.

8.10. A thin uniform circular disk of mass m and radius r has a constant spin $\dot{\psi}$ about a pin at A, as shown in Figure 8.25. The pin twists with the light rigid rod OA of length ℓ at constant rate $\dot{\theta}$ about the horizontal axis (that is, along OA) which is driven about a vertical shaft $B_1 B_2$ at a constant precession $\dot{\phi}$. Note that B_1 and B_2 are the supporting bearings. Disregarding all friction, determine

(a) the kinetic energy of the system for this particular instant and

(b) the total moment exerting on the pin at A.

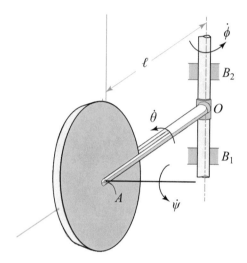

Figure 8.25: Circular disk with a constant spin $\dot{\psi}$.

REFERENCES

[1] Meirovitch, L., *Methods of Analytical Dynamics*, New York, McGraw-Hill, 1970. 145

[2] Roberson, R. E. and Schwertassek, R., *Dynamics of Multibody Systems*, New York, Springer-Verlag, 1988. DOI: 10.1007/978-3-642-86464-3. 145

[3] To, C. W. S., *Introduction to Kinematics and Dynamics of Machinery*, Morgan & Claypool Publishers, 2018. DOI: 10.2200/s00798ed1v01y201709mec007. 148

Author's Biography

CHO W.S. TO

Dr. To obtained his doctoral degree in sound and vibration studies from the University of Southampton in April 1980. He is currently a fellow of the American Society of Mechanical Engineers (ASME) and a professor emeritus in the Department of Mechanical and Materials Engineering at the University of Nebraska (UNL). Prior to joining UNL he was a professor (1994–96) and an associate professor (1986–94) at the University of Western Ontario, Canada. He was an associate professor (1985–86) and an assistant professor (1982–85) at the University of Calgary. Between 1982 and 1992 he was a University Research Fellow of the Natural Sciences and Engineering Research Council (NSERC), Canada. He served as chair of the ASME Finite Element Techniques and Computational Technologies Technical Committee. In addition to being a member of the editorial boards for several refereed international journals, Dr. To currently serves as an associate editor of the *International Journal of Mechanics*.

His main academic interests are in nonlinear stochastic structural dynamics, nonlinear finite element analysis with particular reference to laminated composite plates and shells, nonlinear dynamics and control, machinery noise and vibration diagnostics, and mechanics of carbon nano-tubes. He is the author of *Introduction to Kinematics and Dynamics of Machinery* (2018, Morgan & Claypool Publishers), *Introduction to Dynamics and Control in Mechanical Engineering Systems* (2016, ASME and Wiley), *Stochastic Structural Dynamics: Application of Finite Element Methods* (2014, Wiley), *Nonlinear Random Vibration: Analytical Techniques and Applications* (Second Edition, 2012, Taylor and Francis/CRC Press), and *Nonlinear Random Vibration: Computational Methods* (2010, Zip Publishing, Columbus, Ohio).

Index

Printed in the United States
by Baker & Taylor Publisher Services